THE LIVES OF WASPS AND BEES

The Lives of Wasps and Bees

SIR CHRISTOPHER ANDREWES
F.R.S.

1969

CHATTO & WINDUS
LONDON

Published by
Chatto & Windus Ltd.
40 William IV Street
London WC2

*

Clarke, Irwin & Co Ltd.
Toronto

SBN 7011 1433 9

Printed in Great Britain by
Ebenezer Baylis and Son, Ltd.
The Trinity Press, Worcester, and London

Contents

Acknowledgments

The author and publishers would like to thank the following for permission to include photographs: Mr. S. S. Ristich for Plate 1B; Dr. J. Boorman for Plates 9 and 13B; Dr. Howard Ensign Evans and George G. Harrap & Co. Ltd for Plates 3B, 4 and 7B; the Natural History Photographic Agency for Plates 16A and 16B; and Mr. Guenter Olberg for the remaining photographs. They would also like to acknowledge that the following diagrams have been based on those included in the following books: Figs. 1, 2, 3, and 4 from *Wasp Farm* by Howard Ensign Evans published by the Natural History Press; Fig. 6 from *Wasps Social and Solitary* by George W. and Elizabeth Peckham published by Constable & Co. Ltd; Figs. 5, 7, 8, 9, 11, 12 and 14 from *Handbooks for the Identification of British Insects—Hymenoptera, Introduction and Keys to families*, Vol. VI, Part 1 by O. W. Richards published by the Royal Entomological Society, London; Fig. 10 from *Biology of North American Vespine Wasps* by Carl D. Duncan published by Stanford University Press; and Fig. 15 from *The Behaviour and Social Life of Honeybees* by C. R. Ribbands published by the Bee Research Association Ltd.

Plates

Preface

FOR many people there are prospects of increased leisure in the future. Not all will want to watch professional football or play bingo; many will prefer to get out into the country-side, though, unfortunately, they may have to go farther and farther from towns to find it in an unspoiled state. Of those who do get out, some will turn on their transistor-sets or play at ball games with their children. For those who make the effort to leave the beaten track, there are great pleasures in store. Watching birds, finding and naming wild flowers, catching moths and butterflies or rearing their larvae—these are things likely to attract more and more people, and books to help them are readily available. There are, however, also the insects of the so-called 'neglected orders'; about them most people have little knowledge and their remarkable habits are, in general, quite unknown. This book will accordingly try to show that among wasps and bees in particular there are many with fascinating and often bizarre habits; only a little effort is required to learn to recognize the common ones and to learn something about them. A guide to available books, aiding identification, will be found at the end of this volume, under 'Further Reading'.

Wasps and bees are of particular interest because of the light they shed on the evolution of patterns of behaviour. The many solitary hunting wasps are thought to have evolved from parasitic insects such as ichneumon flies and there are clues as to how this may have come about. Then there is evidence that bees have derived from some of these hunting wasps. Still more important and interesting are the wasps and bees with partly social habits, giving clear leads as to the evolution of the complex societies of social wasps and bees with its climax in the hives of the honey bee. A related theme will be the description of the reaction of all these insects to unfamiliar stimuli, with the light it sheds on the nature of 'instinct'.

Most of the facts dealt with concern European and North American wasps and bees, though there are some references to tropical insects also.

It is a pleasure to thank Professor O. W. Richards for helpful advice; also Drs. I. H. H. Yarrow and C. G. Butler, who, between them, read all the manuscript and offered very helpful criticisms.

CHAPTER I

Wasps and Bees Introduced

HONEY-BEES, bumble-bees and social wasps or yellow-jackets are familiar to all; yet many people do not realize that besides these are dozens of species of solitary wasps and bees. These do not have a socially-organized life, but each female makes a burrow or other form of nest for herself and cares only for her own offspring. In Britain there are something like 200 species of solitary wasps and about the same number of solitary bees. In contrast are half a dozen social wasps and about twenty species of social bees; all the latter apart from the honey-bee, are bumble-bees (*Bombus*).

Across the English Channel the insect-fauna is much richer: species of wasps and bees become more numerous the further south towards the Mediterranean one goes. In North America there are very many hundreds of species. Few of these are identical with the European ones but most belong to dominant genera which are well represented on both sides of the Atlantic. There are, in general, close similarities between the habits of species of any one genus, whether in Europe or America; yet many individual species have fascinating traits of their own.

Few of these insects have common names. People less familiar with scientific names are apt to be concerned lest they do not pronounce them correctly. Accordingly, at the end of each chapter is a list of included species with an indication as to acceptable pronunciation. Not all entomologists pronounce Latin names alike; so there is no absolute standard of correctness. The same lists will indicate by letter whether the species occurs in Europe including Britain (B), Europe but not Britain (E), North America (A) or elsewhere. An appendix to the book (p. 193) will list all the genera referred to under families and other categories, so that some idea can be gained of their relationships. The number of possibly unfamiliar words will be kept to a minimum.

Wasps and bees are members of the order Hymenoptera, which means membrane-winged. Some hymenoptera, such as worker ants, are wingless. In the winged species the front and hind wings are joined together by means of small hooks so that each pair functions as a single wing.

A primary division of hymenoptera is into sawflies and the rest. Sawflies, which of course are not true flies at all, have the abdomen joined to the thorax without any constriction. Their larvae are much like those of butterflies and moths, and like them, feed on vegetable matter. Those of other hymenoptera are maggot-like and legless. In all but the sawflies there is a 'wasp-waist', a constriction between thorax and abdomen. A further subdivision separates the *Parasitica* from the *Aculeata*. Within the *Parasitica* are a number of superfamilies, and within them families, of parasitic insects. Most familiar are the ichneumon flies, very slender insects of which the females mostly have long sting-like ovipositors for inserting eggs into their victims. Others are extremely small; the larvae of some develop within the eggs of other insects. One group has very different habits, being responsible for causing galls such as the familiar oak-apples.

The other big division, the Aculeates, is so-called from the needle-like sting of the females. Included among them are the ants, which we are not dealing with, the pompilids, or spider-hunting wasps, the vespids—including the social wasps—, the sphecids or fossorial wasps, some small groups considered in chapter 11 under 'Wandering Wasps' and finally the bees.

One feature of their life-histories will seem strange to anyone more familiar with moths and butterflies. The larvae of those insects when they have finished feeding, turn into a pupa from which in due course the perfect insect hatches out. Full-fed larvae of hymenoptera, on the other hand may, like those of Lepidoptera, spin a cocoon, but they then change into a pre-pupa or resting larvae, not very different from the feeding larva, and they remain like this for some time, only turning into a true pupa perhaps a few weeks before they hatch out.

It is believed that the vegetarian sawflies constitute the oldest, most primitive group of hymenoptera, that the parasitica evolved from them and the wasps in turn sprang from these and the bees from the sphecid wasps.

The parasitic hymenoptera are sometimes called parasitoids, parasite-like creatures. A true parasite lives in or on its host but commonly does not kill it. Larvae of ichneumon-flies living inside their victim, slowly and inevitably consuming it, may be considered as having in a sense the habits of a predator rather than of a parasite. The typical behaviour of solitary wasps is rather different again. These commonly provision their nests with one or more living paralysed victims, deposit their egg, close the nest and leave the larva, when it hatches, to devour the food provided. The wasp's sting, when it paralyses but does not kill, ensures that the prey will remain fresh for some time; the larva, when it eats, will leave the vital organs till the last. A number of modifications of this basic plan will come to light in the following chapters.

Many wasps and bees have adopted yet another mode of existence, a modified parasitism. Instead of making and provisioning their own nests, they enter those of another, often closely related, species and lay their egg on the food stored there—insects or spiders in the case of wasps, nectar and pollen in the case of bees. The parasite's larva commonly hatches before that of the host and destroys it before proceeding to consume the provisions. Insects with this type of behaviour have been called work-parasites, inquilines or cuckoos: I shall use the last term as the behaviour is not unlike that of the infant European cuckoo in the nest of the pipit or hedge-sparrow. (In America it is the cowbirds which act in this manner: American cuckoos have no such habits.)

Fascinating matters which will come to light in the following pages concern how the behaviour of the solitary wasps has evolved from the parasitoid habits of the ichneumons and how the cuckoo-habit has developed in different groups of wasps and bees. Another evolutionary thread to follow will be to trace how the highly successful social wasps and bees have evolved from the solitary ones. These often live in 'colonies' in which many nests are found close together yet the females work only for their own offspring. In the fully-developed social pattern, there are a few egg-laying females, usually only one, the queen, and much larger numbers of workers. These are incompletely developed females and their labour is for the benefit of the queen's offspring, not their own. All sorts of intermediate stages can be traced between the behaviour of the most primitive solitary bee or wasp and that of the social wasps and honey-bees. The paper-wasps *Polistes*, described in chapter 13 and the *Halictus* bees (chapter 18) provide instances of important intermediate stages. A very interesting difference among the solitary wasps is between those which practise mass-provisioning and those which do progressive provisioning, In the former group the mother seals up the egg with all the food supply the larva will ever need: she never sees her offspring except as an egg. In the latter group the mother leaves her nest unsealed and takes successive food supplies to her larva, increasing the food-supply as the larva grows. She thus does have contact with her young. As the insect food provided will be consumed quickly it can be killed: it will not go bad before the larva is ready to eat it. For mass-provisioning it is more necessary to use the paralysing but not too quickly lethal sting.

Our knowledge of the lives of wasps and bees has come from studies of two kinds, observation and experiment. Many patient watchers have studied the behaviour of these insects under natural conditions in the field, often under conditions of the gruelling heat which wasps and bees seem to enjoy. Others have interfered with their doings in hope of learning how the insect's 'mind' works. Jean Henri Fabre, a school-master living in the south of France, was a pioneer, practising both methods. His work is very well known, for his papers were written in a popular style and have been translated into English[6-11]. He added enormously to our knowledge and stimulated interest in the habits of his insects. He had no faith in Darwin's views on evolution as a result of natural selection and indeed crossed swords with that authority. Many of his arguments make strange reading nowadays, when the essence of Darwin's thesis is generally accepted. Later workers have criticized many of Fabre's interpretation of what he saw, for he conceived of the behaviour of his subjects as rather too cut-and-dried.

A number of workers in the United States—the Peckhams[16] in the 1890s and 1900s, the Raus[17] and, more recently K. V. Krombein[13] and H. E. Evans have added much to our know-ledge of solitary wasps. A popularly written account by Evans, 'Wasp Farm'[4], conveys much information in an entertaining manner. A book from Germany by G. Olberg[15] is full of extraordinary pictures of wasps performing a great variety of actions; many of his photographs are, with permission, repro-duced in this book. Other workers, particularly Richards[20], Butler[1], Free[12] in Britain and Michener[14] and Duncan[3] in America, have devoted more attention to social insects. The more important books and articles are listed in an appendix on further reading (p. 189).

Many wasps and some bees, particularly cuckoo-bees, are brightly coloured, often being banded with black and yellow. These are warning colours intended to deter predators by indicating a repulsive taste or possession of a sting. We can infer that this is successful since so many other insects devoid of stings,—flies, especially hover-flies, beetles and even moths —have evolved colour patterns and shapes resembling those of wasps.

Wasps and bees may be relatively free from the attacks of predators, yet they are subject to attack by countless parasites.

The cuckoo-bees and cuckoo-wasps have already been mentioned. Ruby-wasps and velvet-ants (chapter 11) have similar behaviour; and particularly troublesome in this way are a number of flies of several families. Ants often obtain possession of a paralysed insect which a wasp is about to store away. There are parasitic beetles, mites and worms. It often happens that only a small minority of nests will yield the intended perfect wasp or bee: far more will produce some species of cuckoo or parasite. A recurring theme in subsequent chapters will be the means evolved by wasps and bees to thwart these enemies. The interaction of host and parasite has doubtless been one of the main agencies determining the course of evolution in the aculeate field.

CHAPTER 2

The Spider-hunters

THE Pompilids (*Pompilidae*), named after an important genus *Pompilus*, belong to a big family of wasps, well represented all over the northern hemisphere (Plate 1A). All of them provision their nests with spiders, though they are by no means the only wasps to do this. They are slender insects, some very small, a few as big as the social wasps. They have long legs, often furnished with long bristles or spurs or even combs which may be used in sweeping up earth or sand when they burrow. Even more than with other wasps, they are often to be seen rushing about excitedly, quivering their wings in an apparently agitated manner, turning to one side after another in their search for prey, acting as though every second were precious.

Four habits, apart from the use of spiders as prey, are characteristic of the family, though there are exceptions[29]. A single specimen of prey is used to provision a cell; the paralysed spider is transported by being seized in the wasp's jaws and carried backwards along the ground; the female uses the apex of her abdomen as a tool for pounding earth when closing the nest, or as a trowel; and usually, and in contrast to the habits of most other aculeates, the prey is caught before the nest is prepared.

K. Iwata[52] has classified wasp behaviour according to the order in which various manipulations are carried out. The evolutionary aspects of this will be discussed more fully in chapters 15 & 27. Most pompilids belong in the group in which the following sequence occurs: hunt, paralyse, transport, nest-build, oviposit, close nest. There are examples in the family of more primitive behaviour, where some of these stages may have been lost, as in 'parasitic' species.

Some pompilids specialize in attacking a particular species of spider, others are catholic in their tastes and take spiders of a number of different genera. The European *Anoplius fuscus* takes spiders belonging to five different families. Whether they catch one sort or many depends largely on their hunting habits; if these are directed to finding a spider with unusual specialized habits, then, obviously, the wasps will only catch that kind of spider. Most often a particular pompilid species will catch spiders of several sorts all of which have similar habits — for instance either orb-weavers, tunnel-builders or trap-door-makers. Then, the method of attack will vary according to the nature of the spider's retreat. A species of *Pompiloides* was seen by the Raus[17] to enter a spider's web and walk about on it seeking the occupant without becoming entangled itself. Others threaten and terrify their spider till it lets itself fall to the ground; it is then easily stung. It must be a fearsome problem for a wasp to enter the burrow of a tunnel-maker, when the occupant is facing outwards with its poison-fangs at the ready. The spider in some instances may weigh ten times as much as the wasp. Ferton[41],[44] describes how *Pompilus vagans* attacked a spider (*Nemesia*) which had a second entrance to its lair. It went a short way into one entrance, then out and then into the second entrance, confusing and upsetting the spider, which finally ran out, apparently in panic and became an easy prey. Fabre[11] describes similar tactics on the part of *Pompilus apicatus* which attacks a large black cellar spider, *Segestria*, which lives at the bottom of a funnel. Both wasp and spider kept popping in and out of the funnel, threatening each other. The *Pompilus* would catch the spider by the leg, only to lose it again. Finally she dragged the spider out and dropped her. The spider curled into a ball — a method of defence doubtless useful against some enemies; but the wasp then found it a simple matter to administer a *coup de grâce*. Great though the odds against her may seem to be, the wasps nearly but not quite always emerge the victors from these contests. Many will remember the titanic struggle between a large spider and a wasp in Disney's film of life in a desert.

In some instances the spider's lair is an eminently suitable place in which the infant wasp can feed and grow; the mother wasp can then omit the transport stage in the sequence of operations, can lay her egg on the paralysed spider in her own retreat, close the entrance and depart to look for the next victim. This happens with a Californian species which preys on trap-door spiders. Some species of wasps specialize in attacking spiders which have lairs suitable for this sort of treatment; some of these wasps lack and have presumably lost in the course of evolution the combs possessed by their relations which have to excavate holes for their nests. One European genus, *Homonotus*, has gone even further[59]. Its victims are spiders of a genus which spins together leaves to make a nest. The spider is attacked and stung, but recovers and goes about its normal occupation again. The wasp egg hatches, the larva sits on the spider's abdomen, gradually devouring and finally killing it. This is in no way different from the behaviour of some ichneumon flies of which the larvae devour their victim from outside.

When a spider is stung the poison injected may either cause only a light temporary paralysis or one which is profound. The sting in all aculeates has been evolved as a modification of the ovipositor or egg-laying implement of other hymenoptera. The stings of most solitary wasps are hardly painful to a human being: those of pompilids, however, hurt quite a little. As already mentioned, spiders stung by *Homonotus* wasps fully recover—until they are eaten! Observers have on many occasions removed paralysed spiders from nests and observed how long they will survive. The Peckhams[16] found that some were already dead while others survived up to forty days. When a stored spider is lightly paralysed and seems able to move its limbs, this doesn't do it much good as the victim is commonly packed tightly into a cell, waiting for its devourer to get busy. Some pompilid females take a little nourishment from fluid exuding from the puncture-hole made by their stings. They may cut off some or all of the spider's legs to make them easier to transport and more fluid may come from the amputation wounds and be lapped up. Some other kinds of wasps, as we shall see, do more than this and actually consume much of their victims instead of storing them for their offspring.

As a rule the spider is paralysed as a result of a single sting into the underside of the body; one sting is enough, as the nervous centres of a spider are clustered together well forward in the spider's body. The narrow waist of the wasps has doubtless been evolved so that the abdomen can conveniently be bent underneath the victim and the sting applied where it will be most effective. There is one report of a sting being given first in the spider's mouth, where it could presumably paralyse the spider's weapons of offence and defence, and secondly in the usual position beneath. Another species is said to give numerous stings at random and when the spider's movements are sufficiently slowed down, the ventral *coup-de-grâce* is administered.

As already mentioned, most pompilids drag their prey backwards along the ground to a space suitable for digging a nest-hole. Most often the wasp's jaws hold the spider by the base of the hind legs but some grasp the mouth-parts or spinnerets, the organs for producing threads for the web. This method of progress is not satisfactory since the wasp cannot see where it is going. In general, pompilids have found no need to improve upon it, since they provision nests with single specimens of prey, comparable in size with themselves. A few, however, manage to do better. Those of the genus *Dipogon* typically walk sideways and so get something of a forward view. Others manage to walk forwards intermittently. *Pompilus plumbeus* is a grey wasp often seen on coastal sand-dunes; it occurs both in Europe and North America: it is one of the few which always carries its prey forwards, partly straddling the spider and holding its own body high by walking, as it were, on the tips of its toes.

A few carry their prey in flight, though this is often impossible because the burden is too large. They may drag their spider backwards up a plant and then take flight, gradually losing altitude but gaining distance. This whole operation usually has to be done in several short stages. A few pompilids carry their prey in the air but only just above the ground, so that the spider is partly supported from below. There is one remarkable example of this habit. Captured spiders of the large bog-loving genus *Dolomedes* are towed across water by the flying wasps. This feat was recorded in the literature, but for many years the wasp concerned was unknown; finally Dr. H. E. Evans[4] discovered it to be *Anoplius depressipes*. Because they cannot see in front of their course wasps may drag their prey over difficult obstacles, though these could be very easily circumvented. One wasp carried its spider more than 100 yards. When there is no prepared nest, one wonders if the wasp does not, in many instances, carry its prey for quite unnecessarily long distances, One way and another we have to admit that pompilids are not the brightest specimens of the wasp family; they are of the old conservative stock and compared with others we shall meet later are definitely behind the times.

Next, our wasp has to make its nest and at once we perceive a grave disadvantage in the old-fashioned habit of most pompilids of hunting first and nest-building after. For the prey has to be left unguarded while the digging is going on, and there are many enemies only too ready to take advantage of this. The prey may be stolen by another pompilid. The large black-and-red banded *Anoplius fuscus* is very unscrupulous in this respect[15]. A related species, *Anoplius infuscatus*, even opens up another wasp's nest to get hold of a spider, covering it temporarily while it prepares a new hole for it. This would seem to involve more trouble than catching its own spider, but perhaps it is a coward and fears the preliminary battle.

Other enemies are referred to later. A great danger comes from ants. They are often seen running all over the place and may only too easily chance upon a temporarily abandoned paralysed spider. The large five-spotted *Pompilus quinquenotatus* is called by the Peckhams[16] the 'tornado wasp' from its rushing tactics; its habit is to park its spider in a crotch of some plant while it does its digging; it is thus less likely to be found by ants. *Pompilus plumbeus* hides its spider in a shallow grave while it digs a more permanent tomb for it. Many pompilids seem to be nervous about their treasure while they dig and may interrupt this and return several times to their spider to make sure it is still there. If the ants have got at it, it will probably be abandoned. Most wasps, as we shall see later, circle round a nest to imprint its position on their memory. Pompilids do not ordinarily have to keep returning to a nest for further provisioning, so have not had to learn to be very adept at navigation and sometimes, after a visit to their parked spider, they fail to rediscover their half-finished nest. They will then rush wildly round in a great state of excitement and turmoil.

The commonest nest is one dug in earth or sand. The combs of bristles on the legs are used for shovelling, the front legs being used alternately; or jaws may be used as well. One wasp took just over an hour to dig its burrow. Tunnels in fine clay or earth may be straight but in loose sand there is often a short straight passage and then a right-angled bend. The making of a tunnel may be preceded by several false starts, soon abandoned. Perhaps the texture of the earth was unsuitable; it is also suggested that the habit has a sort of psychological basis, the wasp having to be gradually worked up till its 'nesting-drive' is intense enough for the accomplishment of a complete job. There are, however, other nesting habits among the pompilids. The American *Dipogon sayi* nests in various kinds of natural cavities, making compartments arranged in series. A European *Dipogon* has bristly beards on the front of its face and is said to gather spider-webs on these and to use them for making the partitions between the cells[4]. All sorts of other material may be used by other species. A few pompilids build mud cells under logs or elsewhere.

The nest being dug and the spider—with luck—retrieved, it is dragged backwards into the hole, usually held by the spinnerets, and placed in position. The egg is now laid and every wasp has a favourite spot for oviposition. Most pompilids lay their egg on the spider's abdomen, some obliquely, others parallel to the spider's body. *Ceropales*, to be discussed in a moment, conceals its egg in part of the spider's respiratory apparatus, the book lungs.

The burrow is now closed and the female pompilid is characteristically furnished with a suitably adapted apex to her abdomen so that the earth can be hammered down to make a tidy entrance. One wasp is recorded as having continued to pound down the earth for a whole hour. Some wasps disguise the entrance by strewing little stones and so forth around it.

Mention has been made of pompilids which at times steal each others' spiders. Others are invariably parasitic, or rather 'cuckoos'. Species of *Ceropales* are black wasps with red or white markings. Their habit is to watch other pompilids such as *P. plumbeus* at work[15]. Several at a time may hang around watching their industrious victim (Plate 1A); then one dashes in and lays its egg in the book-lungs of the spider as it is being dragged into a hole. The *Pompilus* seems well aware of the danger and offers fight: the *Ceropales* does not resist yet lets slip no chance of inserting its cuckoo-egg. The *Ceropales* egg hatches before that of the *Pompilus*: the larva destroys the *Pompilus* egg and proceeds to consume the spider. Some wasps of the genus *Evagetes* behave in a slightly different way. A blue species, *Evagetes parvus*, rushes into a burrow which its host, an *Anoplius*, is filling with prey, and somehow manages to pass the rightful owner[4]. The latter oviposits, closes the burrow and departs. Meanwhile the *Evagetes* eats the *Anoplius* egg and lays her own and makes her way out, some five minutes after the host-wasp's departure. Several *Pompilus* in Europe have similar habits, living at the expense of other *Pompilus* species: *P. crassicornis* is a red and black species with such habits and is not uncommon in southern England[59]. These wasps seem able to recognize where another pompilid has nested; they then dig down, discover the buried spider and substitute their own egg for that already laid there. Finding the host-wasp's nest may, however, not be easy. Ferton[42] observed *Pompilus crassicornis* excavating several holes in a row, then others in a parallel row—the technique of a creeping barrage. Finally it found the nest it was seeking.

There will be a number of occasions in this book to comment on the stereotyped behaviour of wasps and bees and to consider how far it is instinctive, how far too rigid instinct gets the insect into trouble and how often the instinctive behaviour is modified in response to difficulties. The Peckhams[16] observed a *Pompilus* which had stung an Epeirid spider so large that it was out of the question to transport it; on the other hand a pompilid has been seen carrying a small spider comfortably in its mouth, yet quite unnecessarily following tradition by walking backwards all the time. Ferton[42] waited till an *Anoplius fuscus* had provisioned its nest, then laid a paralysed spider at the entrance for the emerging wasp to find. The wasp ignored this and went off, presumably to hunt in the ordinary way. As we shall see later (p. 39) other wasps behave very differently when a similar experiment is carried out.

Species mentioned

Wasps

Pompilus vagans		E	Pomm'-pill-us vay'-gans
—	apicatus	E	ay-pick-ate'-us
—	plumbeus	B	plum'-bee-us
—	quinquenotatus	A	quin-quee-note-ate'-us
—	crassicornis	B	crassy-korn'-iss

Species mentioned (*cont.*)

Anoplius fuscus	B	Ann-op'-plee-us fuss'-kus
— depressipes	A	dee-press'y-pees
— infuscatus	E	infuss-kate'-us
Pompiloides	A	Pomm-pill-oy'-dees
Homonotus	B	Ho-mo-note'-us
Dipogon sayi	A	Die-poh'-gon say'eye
Ceropales	B	Serr-opp'-a-lees
Evagetes parvus	A	Ev-ajeet'-ees par'vus

Spiders

Epeira	Epp-eye'-ruh
Dolomedes	Doll-oh-meed'-ees
Segestria	See-jest'-ree-uh
Nemesia	Ne-mees'-ee--uh

CHAPTER 3

Grasshopper-hunters

THE big family of sand-wasps, digger-wasps or fossors will occupy the next eight chapters. Scientifically speaking they are the *Sphecidae* or wasps related to *Sphex*, a genus to be dealt with in this chapter. Very many of them dig their nests in sand; sandy commons and sand or gravel pits are excellent places in which to find them. The older English entomologists, catching insects at the end of the last century, did a great deal of their collecting on the sandy commons to the south of London around Woking and Chobham. Some of the rarities they caught there can still be found, but a good many have probably disappeared. Sand-dunes by the coast also provide suitable territories, as do sandy desert areas inland. There are, however, digger wasps which have quite different habits, nesting in holes in wood, in bramble or other stems; so it is not easy to find a really good English word which will cover the whole family.

The wasps of the related genera *Sphex* and *Ammophila* are rather large insects with very thin waists, in some cases very long and slender indeed. There have been some divergences of practice as to which wasps should be included in each of these two genera. It is proposed in this chapter to deal with the larger, more robust, ones which hunt grasshoppers and crickets and are most generally called *Sphex*. Chapter 4 will describe the slenderer caterpillar hunters, *Ammophila* species. Several species of *Sphex* are common conspicuous insects in the U.S.A. and on the European continent; but we have none in Britain. They like warm conditions, as indeed grasshoppers do. There are relatively few species of grasshoppers in Britain compared with the continent and they get fewer and fewer as one goes northwards. The same tendency is seen in the New World. It is even more true of crickets. Grasshoppers and crickets belong to the order of Orthoptera and these are favourite subjects of predation by digger wasps; more than a hundred species of these are hunters of grasshoppers and their relatives.

Nearly all *Sphex* species follow the custom which prevails throughout the digger wasps of digging a nest first and hunting for prey afterwards. This is of course better than the opposite arrangement adopted by most spider-hunters; for the prey has to be left unattended at the mercy of ants and other enemies for a comparatively short time. Fabre[10] studied in the south of France the yellow winged sphex (*S. flavipennis*). This hunts crickets and makes its nest in colonies in bare sand. The first part of its burrow is horizontal and it uses this as a shelter in hard weather. Two, three or four—most often three —cells open into a central passage. Fabre records that this wasp digs with the rakes on its fore legs, making as it does so a shrill singing noise. He could even hear the singing after the wasp had got some way underground. As was mentioned for spider wasps there may be several false starts before the proper tunnel is made; this has been noticed for the common black and yellow American *Sphex ichneumonea*, sometimes called the great golden digger. This wasp takes from fifteen minutes up to more than four hours to dig its hole according to the hardness of the soil[16].

Then off on the hunt! *S. flavipennis* after finding and overpowering a cricket, stings it first in the neck, then between joints of the thorax, finally directing its sting towards the abdomen. Thus, according to Fabre, the three main nervous ganglia are paralysed. Some other authors however tend not to agree with Fabre as to the regularity with which the paralysing poison is injected into exactly those places. Occasionally a grasshopper's big hind jumping legs may be cut off: perhaps the wasp does not trust its poison to keep the victim permanently quiet! As a rule only female grasshoppers or crickets are caught; they are fatter and juicier than males. The wasp, besides stinging its victim, may chew or massage with its jaws some part of the prey, especially the neck where it is exposed between the hard horny coverings of the head and thorax (Plate 2B). This act, called malaxation, may be done without seriously breaking the skin. It seems to damage the nervous centres of the prey beyond what is achieved by the poison of the sting. Also it causes the exudation of fluid of which the wasp proceeds to partake. There is doubt as to which of these results is the real object of the exercise. The act of malaxating the neck of an insect evidently requires very special skill. Fabre tried to copy the wasp and to do it himself, but he only succeeded in killing his grasshoppers. Sometimes the wasps seem to be carried away with enthusiasm and proceed to make a real meal of the prey, which is then discarded instead of being taken to the nest to feed the next generation. It has been suggested that the wasp's infantile appetite is recalled: the meat and juices of the prey are after all what nourished the wasp when it was a larva. Of course normal food of adult wasps consists of nectar from flowers. It is odd that they are carnivorous at one stage of their existence and vegetarian later on. Presumably the protein of their victims is the stuff for body-building and the carbohydrate of nectar a good source of energy for adult activities.

Most species of *Sphex* grasp their grasshopper by its antennae and straddle it with their fairly long legs; they may drag it forward along the ground and it may even be pulled vertically up a wall. Some Sphex species carry their grasshoppers through the air, embracing them with their legs. The grasshopper may weigh considerably more than the wasp does, so that when the wasp reaches its nest it lands with a decided 'plop'. The prey is then laid down close to the nest entrance, the wasp enters, turns round inside, comes out, seizes the grasshopper by the antennae and drags it down backwards. Two grasshoppers are commonly used to provision one cell.

Fabre carried out a number of experiments on his favourite *Sphex* species, one of which caught crickets, another locusts and a third long-horned grasshoppers, a species of *Ephippiger*. He found that these last, paralysed by the Languedocian sphex (*S. occitanica*) remained fresh for a considerable time, up to seventeen days. If he imprisoned live ephippigers, they used up their energy in useless kicking and died in four to five days. He managed, however, to keep one alive for twenty-one days by feeding it with sugar and water.

Another experiment, carried out on the cricket hunter, *S. flavipennis*, is a classical witness to the rigidity sometimes seen in instinctive behaviour. Fabre waited till his *Sphex* had left its cricket just outside its burrow and had entered itself. Before it re-emerged he removed the prey a few inches from the burrow. The *Sphex* clearly disapproved, for it pulled back the cricket to just outside the entrance and went inside again. Fabre again removed the cricket a short distance, with the same result. He was able to repeat this procedure forty times on the same unfortunate Sphex. Is this a record? By no means. Another observer repeated the experiment on *S. ichneumonea* 103 times. It is satisfactory to report that not all Sphex have habits quite so hide-bound. Fabre himself found that in another colony of *S. flavipennis*, the wasps, after two or three experiences of this teasing-removal, learnt to carry their prey straight in, even from several inches away. The Peckhams[16] also found that the *Sphex* they studied learnt, after a few teases, to pull in its prey directly.

Reinhard[18] did a different experiment with *S. pennsylvanica*, the 'Great black wasp of Pennsylvania' which takes long horned grasshoppers, 'katydids'. It places one to six in a cell. When provisioning had begun, Reinhard blocked up three entrance holes, so that when the wasps returned with further specimens of prey they could not take them in. Baffled in this, they dropped their burden and went off for more. After three days, 252 paralysed or dead katydids were found at the entrances to the nests.

Noting that his *Sphex* always pulled grasshoppers in by their antennae, Fabre cut off the antennae from a grasshopper parked outside a burrow. An antenna-less grasshopper was no use to the wasp, which abandoned it and went off to seek a better specimen. Another of Fabre's tricks was to remove the stored prey from a burrow just before it was finally closed. The wasp, doing her tasks in the proper order, proceeded with the closure, unmindful of the fact that the cells were quite empty.

The Raus[17] studied a wasp, *Priononyx atratum*, closely related to *Sphex*. This one, like the Pompilids, caught its prey before digging the nest hole; only in this case the prey was a grasshopper. It left the grasshopper on the ground but was nervous when digging, constantly interrupting the work to see if its prey was all right and each time bringing it a little nearer to the burrow. It was shadowed by some little parasitic flies, apparently *Metopia leucocephala*, a species common to Europe and North America. Species related to this are often called 'miltogrammine' after a related genus *Miltogramma*. The *Metopia* took no action till the burrow was ready to be closed, when they would dash in to lay their eggs. These hatch almost at once. A parasitic wasp, *Stizus*, was seen hunting for a burrow of another *Priononyx* (*P. thomae*). This it promptly entered to lay its cuckoo-egg: too late — the prey was already teeming with fly maggots. The fly had got in first. *Sphex*, when they have caught enough of their prey, proceed to lay their egg; they typically place it cross-wise between the bases of the first and second pairs of the legs. Here there are exposed soft membranes, where the young larva can easily effect an entry and begin to suck the grass-hopper juices. Eggs of these and most other wasps are relatively large objects, several millimetres in length. Some observers think the exact position of the egg is less important than Fabre supposed.

With the egg safely laid, the *Sphex* closes the burrow, hammering down the earth or sand to make a strong door. Unlike the pompilids which use their tails as a hammer, the *Sphex* uses its head, for this purpose. If however, the provisioning is not complete, the nest closure is only temporary, earth being scuffled over 'kicking-dog fashion'.

The egg hatches in two to four days, the first cricket or grasshopper is consumed in six to seven days, the grub moults and starts on the second victim, usually beginning with the belly. The whole of the larval feeding occupies not more than twelve days, often less (Plate 1B). A cocoon is then spun in forty-eight hours, in the case of *S. flavipennis*, in three separate layers, and lined with an impermeable purple shellac. The nymphal state in this species lasts for twenty-four days. When the perfect insect hatches, it stays put in the burrow for three days or so. During this time the pigment develops and the insect, when it digs its way out, is all ready for its free life. This may last up to two months.

There is a genus of wasps, *Isodontia*, closely related to *Sphex* but with very different nesting habits. They make their nests in holes in wood, hollow stems and elsewhere and provision their nests with tree-crickets or grasshoppers. They line their nests with grasses or with fibres from the bark of juniper or red-cedar trees[4]. They have been seen cutting off lengths of grass, perhaps several times as long as themselves and flying off with them. In closing the nests they make a plug of compacted grass, then a section of coiled grass-stems; finally there are blades arranged longitudinally, often projecting from the mouth of the burrow like a broom. This gets gradually worn away till it is flush with the surface.

Some species, instead of making separate cells for each larva, construct a large brood-cell in which a number of larvae grow up amicably together[13]. There is an interesting gradation in behaviour in the genus. *Isodontia elegans* constructs separate cells in the orthodox way, making partitions between them 7 to 20 mm thick. *I. mexicana* only sometimes makes partitions and if it does they are thin ones, 2 to 4 mm thick. *I. auripes* makes a communal brood-chamber, in which two to five larvae reach maturity. Cannibalism only occurs if there is a food shortage.

Besides the little flies already mentioned, many other hazards attend the rearing of the *Sphex*'s young. Birds sometimes attack it as it is entering the nest and rob it of its prey. A parasitic wasp (*Nysson*) afflicts the *S. ichneumonea*, crawling into and investigating any open hole it finds. Moreover two sorts of mutillids, or velvet-ants, (see chapter 11) haunt the colony for similar purposes.

We will close this chapter with something about an African *Sphex* which hunts migratory locusts. As is well known, several different kinds of migratory locusts in different parts of the tropical world suddenly take to mass movements on a colossal scale, laying waste all vegetation wherever they land. Having consumed everything, they fly on to wreak devastation somewhere else. C. B. Williams[26], studying a swarm of Desert locusts in East Africa, observed that the swarm was apparently accompanied by numbers of a large black wasp, *Sphex aegyptia*. It is worth quoting from his account. '. . . a great swarm of Desert Locusts began to settle about 11. a.m.; and within 15 minutes I noticed dozens of a large black Sphecid wasp running about. . . . My two assistants caught 168 specimens within an hour on an area of a few square yards; and there were just as many at the end as there had been at the beginning. In nearly two years previous residence at Amani I had never seen this wasp before.

Immediately on arrival the Sphecids began to burrow, and a little later were dragging paralysed locusts along the ground and into their burrows. This continued all day until dusk and started again the following morning shortly after 7 a.m. Between 1 and 2 p.m. on the second day the locusts began to fly . . . : at 2.15. only 3 live and 1 dead *Sphex* were seen . . . where two hours previously there had been thousands. They departed in such a hurry that they left hundreds of open burrows, many half finished; and paralysed locusts were lying about in dozens.' An observer six miles away said he had seen large numbers of 'black bees'—almost certainly the *Sphex*—in masses 'about the size of a tree', passing low down

1(A) The cuckoo wasp *Ceropales* following its host *Pompilus plumbeus* as it carries its spider-victim to its nest

1(B) Larva of the Great golden digger (*Sphex ichneumonea*) devouring its paralysed grasshopper

2(A) *Ammophila pubescens* bringing a large stone to close its nest

2(B) *Ammophila pubescens* kneading or malaxating the neck of a caterpillar

with a loud buzzing sound. An odd thing is that all captured *Sphex* were females; and Williams wonders if the species was parthenogenetic, that is exhibiting virgin birth—unfertilized females laying fertile eggs. This is a phenomenon to be discussed later (see p. 85). It is well known in the Hymenoptera; usually unfertilized ova produce males, but there are examples where only the female sex is known and where virgin females give rise to more females in indefinite series.

It is amazing that the instinct to follow the swarm of locusts is so overriding that it makes the *Sphex* females interrupt their logical sequence of actions, down tools and be off with the throng. It has been suggested that one advantage gained by insects of various kinds by mass migration is that they escape from specific predators and parasites. In this instance—and there are others like it—the locusts' enemy does not get shaken off.

Species mentioned

Sphex flavipennis	E	Sfex flay-vee-penn'-iss
— occitanica	E	— oxy-tann'-ick-uh
— aegyptia	Africa	— ee-jip'-tee-uh
— ichneumonea	A	— ick-nu-mone'ee—uh
— pennsylvanica	A	— penn-sil-van'-ick-uh
Priononyx atratum	A	Pre-o-nonn'-ix aytray'-tum
— thomae	A	— toe'-mee
Isodontia elegans	A	Eye-so-don'-she-uh elly'-gans
— mexicana	A	— mexi-cayn'-uh
— auripes	A	— ore'-ee-pees
Stizus	A	Sty'-zus

Parasites

Metopia leucocephala	Met-oh'-pee-uh lu-coh-seff'-a-lah
Miltogramma	Mill-toh-gramm'-uh

Prey

Ephippiger	Eff-ip'py-jer

CHAPTER 4

Caterpillar-hunters

WITH wasps of the genus *Ammophila* we reach some with the narrowest waists of all, as Plate 2 shows. Probably they need a very mobile, freely manipulable, abdomen for the elaborate stinging of their caterpillar victims. Before describing their nesting habits, we can refer to observations on their mating behaviour. Fertilization seems to be best carried out on newly emerged females. So as to be all ready for this, male wasps of most solitary species emerge a few days before the females do. *Ammophila pubescens* males have been seen quartering a locality about 2×50 metres in area[38]. Within this area they pounce upon anything which *might* be a female, usually from a distance of about 30 cm. Baerends used wooden models and dead wasps to study what went on. It seemed that the black and red abdomen of the female served as a recognition mark; males may be all black in this species. With some wasps, the bachelor males are so eager to get ahead of their rivals that they even try digging into the soil at places where females are emerging.

Fabre[10] observed that when some *Ammophila* species were digging their burrows in sand, they would remove unwanted rubbish for some distance but would put aside a few chosen stones. Some species carry their rubbish away on foot, while others fly with it and drop it at varying distances from the nest. They may even fly backwards from the nest in order to do this. The sand is excavated by means of the forelegs, which are provided with combs of bristles, and these are used synchronously not alternately as with the pompilids. Before the hunting begins the burrow is closed temporarily, often with a flat stone of appropriate size, which the wasp may have put aside when digging[4]. Baerends saw an *Ammophila* 'trying for size', placing several stones in position before finally picking on one which was just right (Plate 2A). After this, sand may be scuffed over the entrance or objects such as pine-needles left there; these may help the wasp to identify the entrance when she returns from a hunt.

This matter of navigation or orientation is very important for a wasp. Commonly when the digging is finished, she will fly around committing the details to memory. Sometimes digging will be interrupted while an orientating flight is carried out. Some wasps seem to learn after one short circuit, others take longer. It is interesting that the learning of landmarks may be carried out while the wasp is in flight, even though the prey will be brought home by dragging along the ground. Usually a wasp is very clever in finding its burrow, even though this has been covered over by sand; it can, however, only do so within a limited 'home territory'. When taken further afield, it is lost. The Raus[17], studying *Ammophila pictipennis* noted that the earth excavated from a burrow was placed four inches from the hole, so that a returning huntress, if in doubt, could confine its search to exploration four inches from the locating mound. Even so, one wasp took an hour to find the hole it had dug, dragging its caterpillar with it all the time. Thorpe carried out experiments in which he placed metal screens in the path of an *Ammophila pubescens* returning with prey to its nest. The wasp always, unhesitatingly, diverged just enough to get round the obstacle and then resumed its proper course.[2] This species has to be a particularly good navigator, since, as we shall see shortly, it keeps two or three nests going at once, using the method of progressive provisioning.

Though all *Ammophilas* hunt for caterpillars, they have their different tastes. Fabre[10] observed one species which only caught geometer (looper) caterpillars. Wasps of the related genus *Podalonia* are rather stouter, and hairier than *Ammophila*. One of these (*P. hirsuta*) specialized in grey-worms (*Noctua segetum*). These hide underground and feed on roots. The wasp is able to detect their presence by means of some sense-organ, possibly present in the antennae, and proceeds to burrow down and dig them out. Fabre thought the wasp had profound anatomical knowledge and paralysed its prey by systematically stinging it in each of its thirteen segments, or at least in the first nine. The Peckhams[16] do not agree that the stinging is as systematic as that; one wasp they watched stung segments one to seven, another one to four, another nine to thirteen, another stung between segments three and four and then proceeded to malaxation of the larva's neck (Plate 2B). Accordingly it is hardly surprising that the resulting paralysis is of varying degrees of completeness. Some larvae were seen to do a good deal of rolling and wriggling though without dislodging the egg laid on them.

One *Ammophila*, known as *campestris*, was found to hunt for larvae of *Lepidoptera* or for those of sawflies. The latter are, of course, hymenoptera but the larvae are superficially like the caterpillars of moths. The lepidopterist who wishes to breed from the larvae he catches has to count their prolegs, or false legs, to discover which is which. Adriaanse discovered that '*campestris*' in fact consisted of two species, one of which, the true *campestris*, hunted sawfly larvae; the other, called *pubescens*, is the one which takes caterpillars.[25] When suspicion arose because of these differences in habit that two species were concerned, careful study revealed that the habit differences were related to small structural differences. The two species can evidently distinguish by some means the two kinds of larva—but presumably they don't do this by counting their prolegs. An American species, *A. Aztec*, takes sometimes sawfly larvae, sometimes caterpillars, but the nests examined by Evans[36] usually contained larvae of one or other sort; very few had mixtures. Individual wasps usually, but not invariably, provided one kind of diet for their offspring.

All species carry their larvae, ventral side up, embraced by the legs. Some drag along the ground, others carry in flight. The Peckhams[16] watched an *A. urnaria* dragging a larva along the ground for 260 feet, taking more than two hours over the task. It made better time, however, than another species which only achieved a speed of 5 inches a minute. A *pubescens* may fly when carrying smaller larvae, but may have to drag the larger ones: not very surprising.

Two writers, W. M. Wheeler[61] and H. E. Evans[38] have remarked that the genus *Ammophila* is of particular interest for the student of evolution as it contains examples of different sorts of behaviour which can be arranged in a series. There are strictly solitary ones, using a single large caterpillar for each nest; those which mass-provision with several smaller larvae; those in which provisioning is delayed, the last prey being brought in after the egg has hatched; those which regularly employ progressive provisioning; and finally a few species which keep going several nests simultaneously. Particularly interesting are the observations of Baerends on *A. pubescens*; this by the way, is the only non-social wasp in Britain in which progressive provisioning has been observed.

This wasp worked in turn at two or three holes in which the provisioning was at different stages, having been begun at different times, probably on different days. The *Ammophila* visited each nest at the beginning of the day's operations and evidently determined how many caterpillars were needed to satisfy the needs of the various wasp-larvae. According to what it discovered it would devote its attention during that day to stocking up one particular burrow. If before this visit, the observer added to the store of larvae or removed some, the wasp would modify its actions appropriately and food would be brought where it was most needed. If, however, the additions and subtractions were made after the wasp had completed its 'morning rounds', this interference would be ignored and food would be brought according to what the morning inspection had revealed as necessary.

Evans[36] discovered that *A. Aztec* in Wyoming exhibited very similar behaviour; nest-inspection was, however, not only in the early morning. In this wasp, two days were needed for hatching of the egg and about five for larval feeding. The nest was finally closed when the larva had undergone its last moult and was two-thirds grown. All the nests attended to by one female were within an area of something like a square foot.

Ammophila has again something to tell us about the next step, that of nest-closing. Mention has already been made of how the wasp 'tries for size' to obtain a stone suitable for temporary closure of a nest. *A. Aztec* uses three methods. An empty nest may be blocked by a single pebble and a little sand. When the nest has been only partly provisioned, temporary closing is effected by means of a stone of the right size half way down a burrow; other material is packed above it. All the same, this can be cleared away in a few seconds for purposes of entry and even the temporary closure takes only a minute or two. For final closure the blocking stone is placed near the bottom of the burrow and the material in the burrow above is packed down more firmly. Several observers have testified that for achieving a firm closure a number of *Ammophila* species pound with the head as *Sphex* does. But they do more than this: they may grasp a pebble in their jaws and use this as a hammer to consolidate the blocking soil.[16] It has been argued that *Homo sapiens* differs from pre-human primates partly because he has learnt how to use inanimate objects as tools. This may not be strictly true, for chimpanzees and other primates do seem able to use things as tools. But one wonders, if this were a really good criterion, where *Ammophila* would come in! In fact, *Ammophila* is apt to fiddle about with and re-arrange the stones at a shaft-entrance; so the use of a stone as a hammer is not so very different from conventional behaviour. One finds here another example of variability and flexibility in the behaviour of Ammophilas. Some species seem regularly to use 'hammers'. others never, while others sometimes do it and sometimes not.

The varying natural habits of *Ammophila* are so fascinating that it does not seem to have been the subject of many artificial experiments, as *Sphex* has. Ferton,[42] however, records one type of experiment. He waited till an *Ammophila* had finished its provisioning and effected a final closure. He then offered it another paralysed caterpillar. The wasp promptly re-opened its nest but, finding it satisfactorily full, closed it again. The caterpillar was offered once more, with the same result but a third offer was rejected and the wasp flew off.

Fabre[11] did some experiments with very surprising results. Though all Ammophilas take only caterpillars in the normal course, he was able to rear *A. holosericea* on a diet of small spiders and to persuade a *Podalonia hirsuta* larvae to eat an adult black cricket. Experiments on similar lines will be recorded for *Bembix* in chapter 7.

One has to conclude that the extreme specificity of the diet of many wasps is something determined at the level of the hunting habits of the adult, rather than being concerned with the preferences of the larvae. Something rather similar is met with in the very different field of the viruses. Many viruses are pathogenic for particular species of animal hosts, but if by some laboratory trick they can be insinuated into cells of quite a strange species, they will multiply perfectly well. Again the specificity is determined by something operating only at an early stage of the proceedings.

Among the Ammophilas are species which do not normally sleep in the burrows; these are closed from outside, so they have to 'sleep out'. They may be found clinging to grass stems by their jaws and the extraordinary thing is that they are found projecting rigidly at right angles to the stem. The why and the how seem equally mysterious. A few species, for example *P. hirsuta*, emerge in the autumn, hibernate and start work in the early spring. Fabre[10] found them hibernating in banks of sand, emerging to sun themselves on warm days. Much more surprising, he found on Mount Ventoux a collection of some hundreds of this normally solitary wasp sheltering under a broad flat stone. He wondered whether they might not be resting migrants.

Species mentioned

Podalonia hirsuta	B	Podd-alone'-ee-uh her-suit'-uh
Ammophila pubescens	B	Amm-off'-ill-uh pu-bess'-senz
— campestris	BE	— kam-pest'-riss
— aztec	A	— azz'-tek
— urnaria	A	— err-nayr'-ee-uh
— holosericea	A	— hollow-see-riss'-ee-uh

Prey

Noctua segetum	Nock'-tu-uh sedg'-it-um

The Mud-daubers

MANY active professional and other men suffer from acute boredom and frustration when they have to retire. An American professor, G. D. Shafer,[22] was saved from inactivity when he retired by studying the habits of a black and yellow mud-daubing wasp, *Sceliphron cementarium* (Plate 3A). He finally wrote a book entirely about this one species of wasp.

The name 'mud-dauber' comes from its nest building habits. Nests are built under bridges, in barns and against houses, little cylindrical cells of mud being built side by side till they make a mass the size of a fist. Little lumps of mud are collected and each is spread into the form of a ring, the thicker part of the ring being alternately to the right and the left so that the appearance is as in the plate. About forty mud-loads may be needed to construct one cell. Some of these are vertical, some horizontal, but where this is possible they are placed parallel to each other. The wasp may need also to carry in its mouth loads of water so that the mud can be brought to the right consistency. Of course the mud soon dries and the construction becomes a very solid one.

Mud-daubers and *Trypoxylon* species, considered in the next chapter, all provision their nests with small spiders. They belong to the *Sphecidae* and differ from the pompilids, both in structure and habits, not least in storing numerous small spiders instead of one large one in each cell. Their hunting must accordingly be a less exciting and dangerous enterprise than it is for the pompilids. Shafer records that the spider is held belly down, so that its fangs are powerless; the slender waist of the wasp enables it to bring its sting to bear on the underside of the spider. The Raus[17] saw mud-daubers roll their spider into a ball, sting it in a random manner and then administer a more scientific sting after they had brought it to the nest. There may be up to twenty small spiders in one cell. Among those collected are the dangerously poisonous black widow spiders, though some wasps prefer the little sideways-moving crab spiders which often wait for their prey on flowers.

If the cell cannot be finished and provisioned right away, the wasp effects a temporary closure with a sort of mud curtain. Opening up again is a dusty business, for the 'curtain' will have quickly dried: to dispose of the dust the wasp's fore-legs are provided with eye-brushes as well as with the antenna-cleaning device which is found on the forelegs of most aculeates.

Shafer found that some of his wasps lived for three months during the period between May and October. As they got older, their activities became slowed down and the wasps needed more rest-periods; finally their activities failed altogether. He was able to 'tame' a few wasps, so that they would come and take honey from the end of his finger. One 'tamed' wasp was caught for examination and marking; she then became shy and would not take honey for three weeks; after that time, however, she 'thawed'! One particular wasp which he called 'crumple-wing' was unable to fly. So Shafer provided it with a box of mud and fed it; and after three weeks or so the wasp, though unmated, began to build. She was then supplied with spiders, obtained by robbing another wasp's nest. At first she chewed and sucked them, possibly having need of the protein in their juices, but then she began storing them, placing eight in a first cell and eighteen in a second one. Shafer was uncertain whether any eggs were actually laid.

When the wasp larva hatches out it is a feeble thing and sucks the juices of the first two spiders it encounters. Then its jaws are stronger and it is able to chew up the remainder of the provisions, legs and all. It is mature in three weeks and then spins its cocoon, first making loops of silk attached to the cell wall: the cocoon proper is attached to these but with an air space left between itself and the wall outside. The larva within does not turn into a pupa till the following spring. The thin waist does not become apparent until the wasp actually emerges.

Shafer was particularly interested in a matter we have not yet touched upon—how the wasp larva disposes of its waste products; for, remarkable as it may seem, it does not void either solid or liquid matter during the whole period of its growth. In animals generally, the digestive tract or intestinal canal develops first as a tube blind at each end. Then at each end of the animal an infolding forms a little pocket, and these pockets join up with the main tube, forming, at the head end, the mouth and pharynx and at the tail end, the anus. In the wasp, the front end joins up, as in vertebrates, to form the mouth: they could not very well eat otherwise. But at the hinder end, no such joining up occurs until after the end of the feeding period. Until then, all the solid waste is retained inside the larva. In constructing its cocoon the mud-dauber larva makes a large compartment for itself and a smaller one behind; this Shafer calls the 'chuck-chamber'. When all is ready, the posterior infolding joins up with the hind end of the gut, all the faecal matter is voided at once and pushed into the chuck-chamber, a drop of fluid is excreted: this hardens and seals off the chuck-chamber, leaving the main cavity clean and perfectly sanitary. Other wasps have arrangements of a similar kind.

So far, so good; but there are other waste-products of the body's chemical activities, disposed of in mammals by the kidneys and excreted in the urine. The wasp larva has nothing in the way of kidney tissue; the body wastes form a substance called allantoin and this is converted by an enzyme into insoluble uric acid. During the later stages of the creature's development, organs develop which are capable of excreting allantoin. The enzyme which earlier turned allantoin into uric acid now goes into reverse and turns the uric acid deposits back into soluble allantoin and this can now be excreted; it goes out into the gut. Here the absorption of water makes it go back into uric acid for a second time. After the wasp emerges from the pupa it voids this in the form of white pellets and these may be found in old cocoons.

This mud-dauber is common in the United States; and there is another common one, the blue *Chalybion caeruleum*. This has habits very similar to those of *Sceliphron*. The Raus[17] observed it apparently getting entangled in a spider's web: but when the spider came out, the wasp had no difficulty in stinging it and then freeing itself from the web. This manœuvre may not be successful every time, for occasional dead specimens of the wasp were found in spiders' webs. Several observers have seen signs of what might be incipient parasitism.

The blue mud-dauber has been seen occupying abandoned nests of the black-and-yellow *Sceliphron*. It may, however, go further and open *Sceliphron* nests, ejecting the contents and appropriating the nest for itself.

These wasps are evidently tough insects. A *Chalybion* was alive and moving its limbs twelve hours after being beheaded:[17] and a *Sceliphron* was visiting flowers for refreshment, though it had lost all but the very base of its abdomen.

Mud-daubers occur also in Europe, though they are among the interesting insects which do not seem to be able to stand the climate of Britain.

Species mentioned

Sceliphron cementarium	A	Skell′-if-ron see-ment-ar′y-um
— destillatorium	E	— dest-ill-a-tor′-e-um
Chalybion caeruleum	A	Kal-ibby′-on see-rule′-ee-um

CHAPTER 6

Some Wood-borers

MANY wasps make their homes in holes in wood. These they may excavate for themselves; more commonly, however, they make use of abandoned borings of beetles or other naturally occurring holes. Others use the hollow straws of thatch on hay-stacks or excavate their own tunnels in stems of bramble and other plants. While other investigators have specially studied the wasps which tunnel in earth and sand, Dr. H. V. Krombein of the Smithsonian Institute in Washington has specialized in those which use holes in wood and elsewhere, and has written a book about the discoveries he has made by 'trapnesting wasps and bees'.[13] He made his trap-nests by boring holes in small blocks of wood and placing them in likely places to attract his subjects. So as to cater for all tastes, he made his borings of various sizes, mostly 4.8 or 6.4 mm in diameter, with some smaller (3.2 mm) or larger (12.7 mm). Trap-nests were put out in various parts of the U.S.A. Many he could observe himself; others which had been used by various wasps and bees were collected for him at the end of the summer and sent to Washington. He could discover what use had been made of the nests and what had happened to the user by splitting open the wood and examining the series of cells of the nests within. Then the pieces of wood could be put together again and fixed with elastic bands, and a net attached to the end to catch the insects when they emerged in due course. Full-fed larvae would commonly remain indefinitely in a resting state and they therefore had to be stimulated to pupate and complete their development by exposure to a period of cold.

By examining his nests he was able to find out many things about the occupants, and the facts about scores of different

species are recorded in his book. These facts concern the
architecture of the nests, the materials used to make partitions
and perhaps to line the cells, the prey with which the nests
were stored, the life histories of the occupants—that is the
duration of the stages of egg, larva, resting larva and pupa—
and the incidence of attack by parasites. A surprising result
was the frequency with which a nest would be started by one
species and then taken over by another, so that it would perhaps contain a few cells occupied by bee-larvae and others by
wasp-larvae; and very likely a few parasites would be present
too.

Among the wasps which Dr. Krombein studied were the
very slender species of the closely related genera *Trypargilum*
and *Trypoxlon*. The wasps of the former genus are mainly
North American; the latter has representation in Europe,
including three species in Britain. They are more than wasp-
waisted, for they are altogether extremely slim insects; this
is doubtless an adaptation to the habit, which many of them
have, of nesting in narrow holes in old wood, in twigs, or in
mortar; they prefer holes already made for them by some
other insect such as a wood-boring beetle. There is dispute
as to whether they ever dig out their own holes. Several
observers claim to have seen them doing this. Ferton watched
one leaving its burrow and flying backwards for a few inches
before dropping a load of 'sawdust'. Krombein, however,
thinks these observers may have been misled and that the
wasps were only removing debris left by a beetle which had
previously occupied the tunnel.

The burrows if too wide may be narrowed by lining with
mud. The cells, of varying number, are built in series. There
is usually a barrier of mud between each cell, as in the common
European *Trypoxylon figulus*, but pith may be used. As a rule
there is a double barrier at the entrance to the burrow; that
is there is a final cell which is left empty. One, the largest
American species, *Trypargilum politum*, builds and provisions
a series of cells in a tube of clay, then other tubes beside it:

this insect is called, from the nature of the nest, the pipe-organ wasp.[4]

Spiders of various families, usually small specimens, are caught. Some species seem specially fond of orb-weavers of the genus *Epeira*. Hamm and Richards[50] in Britain watched *T. clavicerum* hovering in front of the web of a young garden spider, *Epeira diadema*. 'The spider moved so as to face the wasp, which was evidently recognized as an enemy.' Several *Trypargilum* species have been seen making dashes at a web and frightening the spider into dropping on to the ground. They then easily catch and sting it. Though several wasps of this genus may be found in one locality, their habits are slightly different. Evans[4] writes of the 'CCCC—principle': complete competitors cannot co-exist. If two insects occupy a precisely similar 'ecological niche', one, it appears, must be slightly more efficient than the other and therefore bound, in time, to push it out. So it was found that *T. tridentatum* likes rather more open country than other *Trypargilum* species. Some species take predominantly snare-building and others prefer wandering species of spider. *T. tridentatum*, like *Sceliphron*, has been found hunting the poisonous black widow spiders, *Latrodectus*. There are also differences between their times of emergence. Different species may store anything from two to twenty-five spiders in a cell. *T. rubrocinctum*, the smallest species, packs in sixteen on the average. There is thus a striking difference from what we saw among the pompilids, all of which store a single, relatively large spider. The *Trypoxylons* are not likely, therefore, to be direct competitors of them.

The Peckhams watched a straw-stack with a surface 20 × 12 feet. Many *Trypoxylon* often nest close together and on this straw-stack twenty were at work. Each of them stored about thirty cells in the course of six weeks of observation; the number of spiders per cell averaged twenty. This means that about 12,000 local spiders met their end. Thirty or so eggs laid by one wasp is a good deal above the average for solitary wasps. The eggs are really quite large in relation to the size of the prey. When laid, as they are, on the spider's abdomen parallel to its long axis, the end of the egg may project beyond the hind end of the spider.

An interesting observation was made on the results of removing the partition between two adjacent cells of *Trypoxylon figulus*: the smaller larva of the two thus brought together always devoured the larger one, ignoring the store of spiders. It thus gets its spider-food predigested!

Needless to say, these, like other wasps, have their enemies, and many parasites are recorded. They have, however, evolved one method of defence which is very unusual among solitary wasps. The males do not all adopt the general habit of idling away their lives on flower-heads; they may take part in guarding nests, when the female is out hunting. When she returns usually after a hunt of ten to twenty minutes they politely make way for her to enter or may climb on her back and enter with her.[16] They don't hunt themselves but may help her in arranging the spiders in the cell, doing in fact the odd job about the house. One male was seen guarding two burrows alternately.

This helpful habit of the males seems to occur only among the American *Trypargilum* species, not in *Trypoxylon*. In any case the males are not very efficient, but in fact rather cowardly and useless in thwarting the sneaking miltogrammine flies or the egg-shooting bee-flies. Some parasites doubtless break in after the burrow has been sealed off. One of the commonest in Europe is the brilliant metallic blue-green *Chrysis cyanea* (see chapter 11). Parasites of the three British species include seven other species of Chrysids, nine kinds of Ichneumon, three parasitic flies and a beetle. A fly (*Amobia*) was seen to settle near the mouth of a burrow and wait till the female came out. The fly instantly entered and in two seconds had deposited a living larva; this is one of the flies which virtually skip the egg-stage, the larva being deposited all ready to go into action.

3(A) The mud-dauber *Sceliphron destillatorium* by its nest

3(B) An aggregation of sleeping *Steniola obliqua*

4(A) *Bembix pruinosa* carrying a horse-fly into its nest

4(B) *Bembix* larva in its tunnel with flies provided in a row

Species mentioned

Trypoxylon figulus	B	Try-pox′y-lon fig′-u-luss
— clavicerum	B	— clay-viss′-er-um
Trypargilum politum	A	Try-par-gile′-um poh-lite′-um
— tridentatum	A	— trydent-ate′-um
— rubrocinctum	A	— ru-bro-sink′-tum

Prey

Epeira diadema	Epp-eye′-ruh die-a-deem′-uh
Latrodectus	Latt-roh-deck′-tus

Parasites

Chrysis cyanea	Cri′-sis si-ay′-nee-uh
Amobia	A-moh′-bee-uh

CHAPTER 7

Bembix and Progressive Provisioning

THE genus *Bembix*—called *Bembex* by Fabre and other early writers—is an important one, with many representatives in North America and Europe. It is another one which, unfortunately, does not reach Britain, at best just reaching the Channel islands. This and some small related genera afford an example *par excellence* among solitary wasps of progressive provisioning, of which we saw an example in the chapter on *Ammophila*.

Bembix wasps are rather stout with relatively large heads, especially large eyes and a conical abdomen without much in the way of a wasp-waist: this is seen in Plate 4A. They exhibit among them a number of interesting features, apart from those involving nest-provisioning.

Males, as is usual among wasps, emerge before females and in some species including *Bembicinus* engage in intricate sun-dances in the air. Such dancing is very common among flies, but rarer in the hymenoptera. Such throngs of dancing males commonly serve to attract a female, who, when she flies among them, acquires a mate in a very few moments. The Raus[17] record that *Bembicinus nubilipennis* nested in large colonies in the hard ground of a baseball diamond. Emergence of the wasps seemed to be simultaneous, though not always at exactly the same time each year. A virgin female would be quickly surrounded by swarms of males and the fortunate suitor would have to be prompt in clearing off with his bride away from the throng. Rau complains that his researches were apt to be interrupted when people wanted to play baseball! The sun-dances of *B. spinolae* are said to produce a hum audible at some distance. Males of *Microbembex monodonta* are so alert that emerging females are usually mated with before they ever have a chance to fly.[5]

Much of what follows in the account of various Bembix species will be seen to be concerned with the need to thwart various parasites. Most troublesome among these are the little grey flies of the genera *Miltogramma* and *Senotainia* (Plate 7B), small relatives of blue-bottles and flesh-flies (*Sarcophaga*). As mentioned earlier, they operate by dashing into an open burrow and rapidly depositing eggs or tiny larvae. Then there are flies related to the bee-flies, *Bombylius* and *Anthrax* (see p. 141) which hover around and hurl their eggs towards holes; the young larvae crawl in when they hatch. The velvet-ants, which are wasps related to *Mutilla* (see p. 85), and the ruby wasps or Chrysids (p. 79) have various methods of attack, perhaps burrowing down and laying their eggs on the fully-fed wasp larva. (see chapter 11). There is even a fly, *Physocephala*, itself rather wasp-like, which attacks and oviposits on a *Bembix* in flight, its larva developing at the expense of the adult wasp.

Bembix species mostly dig their burrows in sand; some like it hard, others burrow into quite loose sand. One species in Texas nests in almost vertical banks, too steep for the wingless velvet-ants to climb up.[51] There are often very many nests together in a colony: each female, however, works only for her own brood. Tunnels may be up to 42 cm long. An interesting feature seen in several *Bembix* species as well as in some other wasps is the digging of accessory burrows.[37] These may be clusters of holes, left unclosed near the true burrow. These may be merely temporary sleeping quarters or they may be there as a tice, to persuade a parasite to lay her eggs in an empty hole or at least to make her waste her time in exploring it. Evans has published a picture of a velvet-ant walking from one open 'false' burrow to another one, and actually passing over and ignoring the true burrow which had been closed and camouflaged. Some wasps when they finally close their true burrow, require material to fill it up and this they may obtain by quarrying and thus making another hole near their burrow. This may have been the origin of accessory burrows, but they have certainly acquired a secondary purpose as indicated above, for they are sometimes made before the true burrow is closed.

Bembix pruinosa makes a special kind of false burrow. A burrow is first dug parallel to the surface and not very deep, then after perhaps 20 inches it suddenly turns and goes in deeply for a similar distance, the whole burrow being as much as a yard and a half long. When the cell has been finished and the egg laid in the empty cell, the mother leaves by a fresh entrance near the point where the burrow began to go down

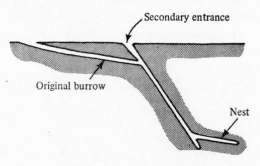

Secondary entrance

Original burrow

Nest

1. False burrow of *Bembix pruinosa*

steeply; the original shallow burrow is filled in (Fig. 1). Provisioning then begins. But instead of the usual cell, there has been constructed a long chamber with the flies lying in a row along it (Plate 4B). The larva progresses along the row, devouring the juicy bits and leaving the debris behind, Evans writes 'it is a temptation to say that the larva is fed cafeteria style'. The mother wasp is not content with mere provision of food; she collects all the debris and walls it off at the far end of the chamber. A likely reason is that fly larvae which despite precautions, gain access to the nest, subsist partly on such debris and partly on freshly provided food. If any or all of them can be walled in, they are safely out of harm's way! In this and other *Bembix* species, the presence of fly-maggot's does not necessarily involve the death of the wasp larva: both may survive. But if the mother wasp has to keep bringing in enough food for both her own offspring *and* the intruders, she will be kept very busy and will probably end by provisioning fewer nests.

One species (*Bembix U-scripta*) has an odd habit, perhaps concerned also with thwarting parasites. Digging her nest results in the appearance of a mound of earth behind the entrance hole. When the burrow is finished, the wasp proceeds to shovel the mound of earth away from the hole, facing forwards and scraping back, till she ends up with a mound some 8–12 cm distant from the hole. Near the entrance itself, the ground is nice and level. Curiously, wasps of this species vary in the extent to which they carry out levelling operations. With this and other *Bembix* species, the wasp commonly rests inside her burrow, effecting a temporary closure from inside. *Bembix sayi*, when it finally closes the nest, packs sand down firmly with the tip of the abdomen and then makes zig-zag or wavy scraping trips away from the entrance to smooth the soil, doing this as many as eighty times. Some American *Bembix* only make one cell in each burrow, others two, others up to six.

To make sure of finding their own nests again, wasps have, as mentioned earlier, to make orientating flights, to familiarize themselves with the surroundings of their nests. Some *Bembix* are evidently very proficient at this, as they come with their prey apparently from a distance, and dive straight into their burrows. *Microbembex* seems to learn her lesson with hardly anything in the way of orientating flights and she was put out very little when an observer moved objects near the nest. When a foot was placed over her hole she just waited till the foot was removed and then dived in. It is possible that the landmarks which guided her were rather distant ones on the horizon of her field of vision.

Most *Bembix* species hunt flies, some being satisfied with flies of any family, others preferring particular kinds of flies: one species for instance specializes in biting horse-flies

(*Tabanids*) (Plate 4A). A European species concentrates on bee-flies (*Bombylius*) and their relatives; this would seem an excellent habit, for the bee-flies are parasites of bees and wasps: so they get food for their young and remove potential enemies at the same time. The American *B. troglodytes* does much the same, but other species merely chase dipterous parasites away and do not try to catch them.[51] These wasps, when seeking prey, commonly patrol likely places such as dung-heaps; they like to pounce on their prey on the wing. The flies, destined to be brought into the nest in instalments to feed the growing wasp-larva, do not have to be kept fresh by paralysing; so they are usually killed. Those brought in later in the course of provisioning may, however, be paralysed. In flies first brought in, a wing, a leg or one side may be removed, so that the fly lies more 'comfortably' on its back. When the larva is small, only a few flies are needed; the daily supply is increased as it grows. One *Bembix* collected altogether more than seventy-nine house-flies (*Musca*), sixty to eighty-two flesh-flies (*Sarcophaga*) and twenty Tabanids. One larva, already starting to spin its cocoon, consumed an additional twenty-six flies when these were offered. Later in its development, a daily ration of twenty flies is not unusual. Feeding goes on for seven to ten days.

There are *Bembix* species which hunt for bugs[51] or for beetle larvae. Fabre[11] persuaded the larva of a species which normally ate flies, to feed on grasshoppers or praying mantis: in fact one larva preferred a mantis diet to one of flies (see also p. 39).

Captured flies are usually carried to the nest embraced by the wasp's middle legs, the fly being upside-down. The egg may be laid on the first fly to be brought in; in other species it is on a fly later in the series. In yet others the egg is laid on the floor of the cell and flies are only brought in as the egg is just hatching. It may be placed on a little pedestal of a grain of sand and this grain may be carried around with the larva for some while. At other times, when the egg is laid on the first fly, that fly, dried up and unpalatable, may serve simply as a pedestal and not as food.

While the various *Bembix* species differ in the various minor ways we have considered, there are others which have departed rather farther from orthodox *Bembix* behaviour. One species, *B. Linei*, nesting on the sea-shore, finds conditions suitable for it only for limited periods, and, having to work quickly, has abandoned the habit of progressive provisioning. Another species, on the other hand, shows what may be taken as an advance, in that it cares for more than one larva at a time. It will be recalled that a few *Ammophila* species do the same. (p. 37).

A species marked with a U, *Bembix U-scripta*,[31] has developed very unusual habits; it does its hunting at dusk when flies are no longer in evidence, having gone to rest for the night. Evans records that though nests were usually full of fresh flies, he could never see the wasps at work doing the provisioning. Usually he made his observations not later than the afternoon, but one day 'we noted a few females provisioning their nests about 1600 hours and remained in the area until 1915. We discovered that there was a fresh emergence of females from their nests just before sunset.' It then appeared that the hunting habits of these wasps differed from the usual pattern: they sought flies, mostly selecting small ones, found nesting on vegetation. At one place in Texas flies found in their burrows comprised thirty-three species, distributed among no less than nine different families. This wasp was evidently doing well for itself, having carved out a very special niche in which it had no competitors.

Another of the Bembix tribe, *Microbembex monodonta*, has apparently been even more successful, for it is perhaps the commonest solitary wasp in many parts of North America.[11, 30]

This wasp has become a scavenger, collecting dead or helpless insects or even bits of insects. It is more catholic in its tastes than any other known solitary wasp, using insects of at least ten different orders. It cruises around, looking for provender for its family, picking things up here and there, rejecting what is unsuitable. Objects collected included dead spiders, grasshopper legs, dead beetles, flies, bugs, moths, caterpillars and, very commonly, worker ants. Curiously enough the wasp has not lost the instinct of her ancestors, for she routinely curves round her abdomen and 'stings', or appears to sting, all her captures, however dead and unsuitable for such treatment. It seems fairly safe to assume, quite apart from the presence of this rather ridiculous 'stinging' habit, that *Microbembex* has evolved from the regular *Bembix* stock, for in other ways habits are similar to those of many other *Bembix*. The prey is carried with the middle legs, the egg is laid in an empty cell, provisioning is progressive, there are many accessory burrows, apparently for sleeping in, and so on. Like other *Bembix* they feed on flowers, and the females steal from each other and quarrel, though never too seriously.

The *Bembix* genus, looked at as a whole, shares something with *Ammophila*. Both genera show such an interesting diversity of habits that they must give the thoughtful observer ideas about evolution. In particular, we see in both groups of wasps the first signs of progress towards the social life we shall see later in *Polistes* and *Vespa*.

Species mentioned

Bembix spinolae	A	Bem'-bix spin-oh'-lee
— pruinosa	A	— pru-in-oh'-suh
— U—scripta	A	— U—script'-uh
— sayi	A	— say'-eye
— troglodytes	A	— trog-loh-dite'-ees
— linei	A	— line'-ee-eye
Bembicinus nubilipennis	A	Bem-biss-ine'-us nu'-billy-penn'-iss

Species mentioned (*cont.*)

Microbembex monodonta	A	My-croh-bem'-bex mon-oh-dont'-uh

Parasites

Senotainia	Seen-oh-tay'-nee-uh
Physocephala	Fie-soh-seff'-a-luh

Prey

Bombylius	Bom-billy'-us
Musca	Muss'-cuh
Sarcophaga	Sar-coff'-a-guh

CHAPTER 8

Bee-wolves

WASPS of the genus *Philanthus* have been favourite subjects for study. They are conspicuous black and yellow insects, some of them fairly large, and they have many interesting habits; in particular, they prey upon bees (Plate 5A). *Philanthus* means flower-loving. They visit flowers for honey as do many other wasps, but also because it is on flowers that they catch many of their bee-victims. *Philanthus triangulum* is the European slaughterer of honey-bees (*Apis.*). It was formerly common in the Isle of Wight and occurred sparingly in other parts of the south of England, but it may now be extinct there. While British entomologists may lament this, bee-keepers perhaps bear the loss rather easily! In America north of Mexico there are no less than twenty-five *Philanthus* species.[40] Two of these specialize in taking honey-bees, though such bees are not native American insects. Most of them are content with smaller prey, taking especially bees related to the genus *Halictus*, (see chapter 18). There are also wasps of the related genus *Cerceris* which hunt bees, and one of them is not uncommon in Britain: we shall deal with these presently.

Perhaps because suitable nesting sites are few, it is common to find a number of *Philanthus* nests close together. Nests are often dug in sand. An American wasp, *P. politus*, one of the smaller species, nests on fairly flat ground, and there Evans watched it digging. Excavated earth was periodically pushed out of the burrow, making a mound outside. At intervals the wasp levelled this off, backing out and kicking the sand aside till a conspicuous sand-heap was no longer visible. Other *Philanthus* species do not trouble to do this, as they like to make their holes in fairly steep banks and the earth or sand they

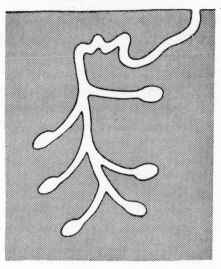

2. Section through burrow of *Philanthus solivagus*

shovel out rolls away spontaneously. All the species make a main burrow with a number of cells approached by side passages: the arrangements vary from one species to another. The large wasp *P. solivagus* makes a tunnel along a zig-zag course (Fig. 2): presumably this makes it harder for a parasite to find its way to the cells. *Philanthus lepidus* on the other hand, makes false burrows (see p. 51) and is the only *Philanthus* known to do this.[34] One mound outside a burrow had no less than five false burrows, all more conspicuous than the true one—enough, one would think, to foil any parasite. Yet the *Senotainia* flies manage to thrive at the expense of various *Philanthus* despite the many dodges for deterring them. They diligently follow wasps returning with prey, flying behind and a little below. The wasp may sometimes get rid of them by flying rapidly some distance off and then returning. Possibly false burrows are more effective against the Mutillids and bee-flies than against the little shadowing flies.

The digging is done with the fore legs, sand flying out backwards between the other pairs of legs. Accumulations may be pushed out with the head as a ram. One *Philanthus* working in rather hard soil took two days to dig a burrow fourteen inches long. Presumably the Florida species which dug a burrow five feet long was working with easier material.

One *Philanthus* weighing 0.031 g shifted 53.6 g of sand. Comparable work by a 150 lb man would involve moving more than 130 tons. The male wasps may dig their own little burrows for shelter, returning nightly to the same bachelor quarters.

Before going hunting, the wasp has to have her geography lesson, and this matter of orientation has been more studied for *Philanthus* than for most wasps, particularly by N. Tinbergen.[24, 25] He tried to find out what it was among landmarks which enabled the European species *P. triangulum*, to find its home so readily. He placed a ring of fir-cones round an entrance-hole and, when the wasp had become accustomed to this, he moved the cones to a circle away to one side. (Fig. 3.) The wasp returned first of all to the middle of the new circle. He then made a circle of alternating discs and hemispheres, and in due course moved them to make a circle of discs on one side and hemispheres on the other: the wasp went

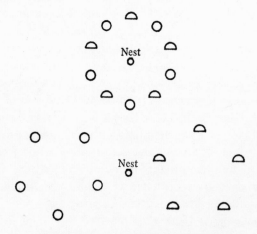

3. Experiment on nest location of *Philanthus*

every time to the hemisphere-pattern rather than to the discs. In similar experiments he showed that other objects sticking up from the ground particularly attracted the wasp's notice: also that a ring with a broken check pattern was more seductive than one of solid black. Tinbergen further made more complicated arrangements of objects round the hole, a block, a little tree, fir-cones, a forked twig. By moving them around he was able to conclude that the *Philanthus* reacted better to the construction as a whole than to any one component. When things were simple, the wasp had only to do one or very few circles to learn its landmarks. If Tinbergen had been changing a number of things to confuse the wasp, it took rather longer over the task. Also, if the wasp had been confined to its hole for a while because of bad weather, it had to make some orientating flights when it came out again to refresh its memory.

Our *Philanthus* has now gone hunting, usually seeking its prey on flowers. The sense of sight operates first in locating the victim. Then, when it gets closer, smell takes over, and if the likely-looking insect turns out to be only a fly, the attack is called off. At times, however, *P. triangulum*, the bee-wolf, makes its pounce from some distance off: it may find it has caught by mistake a drone-fly, *Eristalis*, an insect of about the colour and size and shape of a honey-bee. Then, either touch or smell tells the wasp of its error. Bees may be left momentarily outside the burrow while the wasp looks inside. One observer smartly exchanged a stung bee for a drone-fly stupefied with cyanide. The wasp was never deceived, though it readily accepted stupefied honey-bees. The wasp larva, on the other hand, is by no means so particular. Fabre,[11] in experiments similar to those observed earlier (p. 39) induced *Philanthus* larvae to consume drone-flies. Fabre describes also how the bees are killed. The victim may be gripped belly to belly; the pair then roll over and over until the wasp manages to bend its abdomen forward and sting the bee in the neck. Or else the wasp stands nearly erect, holding the bee with its four front legs while it balances itself on the tips of its wings and its hind legs; it then turns the poor bee round and round till it is in a convenient position for being stung in the neck. This sting seems to be usually fatal. One may wonder why the bee, which also has a powerful sting, does nothing about it. It does seem to try to wield its sting but this is done quite ineffectually. In any case the *Philanthus* is furnished with armour which is not only hard but also smooth and slippery. Having stung its prey, the wasp malaxates the neck with its jaws, and exerts pressure with its legs on the bee's abdomen, to squeeze out the nectar, and this it greedily laps up. One wasp was seen to do this to six bees in succession. Another observer found thirty bee corpses similarly treated outside a nest. Fabre believed that the honey was removed because it would have been harmful to the wasp larvae. He killed some bees and gave them to *Philanthus* larvae without first removing the honey; the wasp larvae perished. The conditions were, however, artificial, and there are several *Philanthus* and *Cerceris* species which do not consume the nectar which the bee is carrying. And anyway there is no doubt that *Philanthus* is very fond of nectar. Fabre watched one continuing to lap up the honey of its *Apis* prey while its own abdomen was being munched by a praying mantis which had caught it.

Most American *Philanthus* catch bees smaller than honey-bees and are especially fond of *Halictus*. Bees of this large genus, to be discussed in chapter 18, have now been distributed into several genera, but it is convenient to use the more familiar name. This does not matter because the wasps catch any bees about the right size. In fact several have been recorded as including small wasps among their victims. *Halictus* species are often caught on the flowers of golden-rod and asters; of these they are specially fond. One *Philanthus* was more enterprising and entered a *Halictus* burrow, stinging the bee in the burrow and carrying it out. This it did four times in ten minutes. Bees of the genus *Sphecodes*, a parasite of *Halictus* (p. 131) and other cuckoo-bees are rarely taken, perhaps because they are less commonly seen on the flowers where the *Philanthus* finds its most profitable hunting.

The prey is carried off in flight; the wasp grasps its antennae and embraces it particularly with the middle legs. It may be dragged for a time: one *P. flavifrons* was seen at the entrance to a bee-hive 'mowing-down' the guard bees and dragging off a victim too heavy to fly away with. The wasp may descend to its hole from a height and dive straight in (Plate 5A). Some species make temporary closures of cells when they are out hunting. Then, when they land, they pass the prey back to the hinder legs, to free the front legs for scraping away the sand at the entrance: then in they go. The bees may be stored one after another in a temporary cache at the bottom of the burrow, perhaps in loose sand. Then the wasp, when she has collected enough, proceeds to store them in a series of cells, laying her egg on the top one in each cell, then closing it. Hunters of the little *Halictus* bees store a number – as many as eighteen at times – in one cell. *P. triangulum*, catching her larger victim, only stores a few: Fabre says there are two for males, more than two for female larvae.

When the larva hatches it does not at first wholly emerge from the egg-shell but sticks its head out and begins its first meal. Later it is able to move around the cell by virtue of a telescopic false-leg at its hind end. Feeding-up is finished in about a week. The larva then spins a thin-walled cocoon suspended from the cell-wall by a network of threads; and so it remains through the winter.

A genus of bees similar to *Philanthus* is called *Cerceris*. These also are striking black-and-yellow wasps. They have hard outer armour but this is not as smooth as in *Philanthus*, often sculptured or covered with minute pits or punctures; it is as if pins of various sizes had pricked them before the armour had set hard (Plate 5B). A number of species prey on bees, while others hunt beetles, especially weevils. *Cerceris rybyensis* is a European species, not uncommon on sandy commons in Britain. I used to watch its colonies on a hard path on

Hampstead Heath in north London. As with some *Philanthus* it mainly catches *Halictus* or *Andrena* bees (see chapters 17, 18), especially the former.[50] It does not bother about the relatively slender males but attacks females laden with pollen. These it knocks down as they are about to enter their burrows. It then sits on the bee's back, holds it by the neck and stings it under the thorax. The bee is then seized by one or both antennae and, supported by the wasp's legs, carried off in the air and taken straight into the burrow (Plate 5B). If it is very heavy transport may be in several stages.

The sting of *Cerceris* is by itself not fatal, but the wasps usually carry out malaxation, and bees do not survive this for long, at most forty-eight hours. In stinging, the wasp slides its sting up the under-belly of the bee till it finds a chink which the sting can penetrate. The malaxation is severe and the skin is broken, for the wasp's tongue can be seen entering a point of penetration. Hamm and Richards[50] consider that the object of the exercise is definitely to feed the wasp rather than to complete the paralysis.

Burrows may be dug by *Cerceris* herself or she may use those

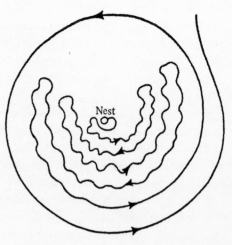

Nest

4. Locality study by *Cerceris*

made by other insects. Miniature mole-hills are left at the
nest entrances and not scraped away. There are up to six or
seven cells off each burrow, and bees stored are either four
to five, or else seven to eight, probably depending, as in the
case of *Philanthus*, on the sex of the offspring. The position
of the nest is memorized by zig-zag flights about a foot above
the ground: the wasp faces the nest while moving gradually
farther and farther away from it (Fig. 4). Consuming its
victim's nectar is not a general habit among the bee-hunting
Cerceris species, though they may lick the bee's tongue. *C.
rybyensis* is said to kill a certain number of bees for personal
consumption.

While we are considering the genus *Cerceris* something must
be said about the more numerous species which hunt for
beetles. Fabre[10] studied several of these. *Cerceris tuberculatus*
was preying on a particular weevil of the genus *Cleonus*; these
weighed rather more than did the wasp. Fabre unkindly
robbed one of its prey as it was entering its nest; this he did
eight times in succession and every time the wasp went off and
quite quickly returned with another weevil of the same
species. Fabre himself, though a knowledgeable entomologist,
did not know where he could have gone to catch that weevil.
Weevils caught by *C. arenaria* and *labiata* would keep fresh
for a month after being stung. Though apparently paralysed
they kept voiding their waste-products for as long as a week.
Fabre suggests that weevils and other beetles hunted by
Cerceris are the only beetles which have their nerve ganglia
close together and so easily paralysed by a single sting. Though
many *Cerceris* prey on beetles and not on bees, Fabre found
that the larvae of *C. arenaria* would readily accept *Halictus*
bees instead of weevils. The Peckhams[16] describe how an
American species of *Cerceris*, when not out on the hunt, sits
with its conspicuous yellow face just inside the burrow, looking as if it were 'leaning on its elbows'. Doubtless this habit
serves to keep out parasites.

Finally, a word about another related wasp, the American

Aphilanthops frigidus. This preys on ant queens,[16] taking its victims when they are winged or after they have shed their wings. They have been seen watching ant-heaps and even entering and being chased out by angry workers. It is suggested however, that the queen ants are usually caught after their descent from a nuptial flight.

Species mentioned

Philanthus triangulum	E	Fill-anth'-us try-an'-gu-lum	
— politus	A	— poh-lite'-us	
— solivagus	A	— soh-lee-vay'-gus	
— lepidus	A	— lepp'y-duss	
— flavifrons	A	— flay'-vee-frons	
Cerceris rybyensis	B	Ser'-ser-iss rib-ee-en'-sis	
— tuberculata	E	— tu-berk-eu-late'-uh	
— arenaria	BA	— arr-ee-nair'-ee-uh	
— labiata	B	— lay-bee-ate'-uh	
Aphilanthops frigidus	A	Ay-fill-anth'-ops fridg'-id-us	

Prey

Halictus	Ha-lick'-tus
Sphecodes	Sfeck-oh'-dees
Andrena	An-dree'-nuh
Eristalis	Erry-stale'-iss

CHAPTER 9

Bug-hunters

'BUG' to a slum-dweller means bed-bug. The term 'bug-hunter' is applied, rather jocularly, to entomologists in general. The word 'bug' has, however, a more scientific meaning, referring to the order *Hemiptera*. This means 'half-winged' and has perhaps most relevance to the first big sub-division of the order, the *Hemiptera heteroptera*. In beetles the first wings have become horny wing-cases or elytra. The *Hemiptera heteroptera* do not go quite as far as that; their fore-wings are horny as to the half nearer the body; the outer half is membranous, as are the hind-wings, and both wings are used in flying. These are the true bugs, often, for obvious reasons, called stink-bugs; and the most conspicuous among them are the fat squat shield-bugs or pentatomids. The second big division of the Hemiptera consists of the *Hemiptera homoptera*, in which the fore-wings may be tougher than the hind-wings but are not divided into hard, membranous portions; they are folded in a roof-like manner when closed. In this group we find the leaf-hoppers, the cicadas, the aphids, or green-flies and the mealy-bugs or coccids. Bugs of all kinds mostly suck plant juices and as you might expect there are wasps which prey upon all the various bug families.

First we will deal with those which hunt for shield-bugs. Conspicuous among them are those of the genus *Astata*, a word meaning restless or never standing still. Many of them are half black and half red (Plate 6A), others all black. Evans[4,30] has described the mating behaviour of the American *Astata unicolor*. The males perch on flowers or on a stone for hours at a time making, every so often, short excursions into the air, up, around and back again. 'Back on his perch, he rotates his body in several directions and walks about a bit before finally coming to rest with his antenna extended rigidly.' The odd thing is that the area frequented by the males is some little way distant from where the females can be found and where they make their burrows. Presumably not-too-bashful virgin females must at times make sallies into the 'male' area.

Nests are dug, often in very hard soil and there are branches from the main burrow leading to the cells which may, some of them, be arrayed in series. The nests are hard to find as they are often under the shelter of drooping foliage. Stink-bug nymphs of several different species are caught; those stung are dead or at least very completely paralysed. Not only does the *Astata* make orientating flights around when she leaves her nest but she also approaches it with her prey in a devious manner, probably in order to deter her enemies. She holds her bug with its antenna in her jaws, straddling it with her legs, at first in the air, then along the ground as the nest is neared. As was described for *Philanthus* and her bees, a number of carcases are stored temporarily before being distributed into the various cells. The egg is laid on the underside of the first bug in each cell. The bugs have tough skins and so the larva hollows each out in turn beginning with the under-belly. The habits of the European *A. boöps* (ox-eyed) are very similar.

Another species was seen being pestered by one of the miltogrammine flies, *Senotainia* (Plate 7B). One was seen flying around for ten minutes vainly trying to shake its enemy off. Wasps may give up in despair and leave their bugs on the ground if they cannot get rid of their tormentors. These, however, too often succeed in depositing a newly hatched larva on the bug before it is taken into the nest. Twenty-six of fifty-eight cells of *A. occidentalis* examined were found to contain the fly-maggots.

A number of other wasps provision their nests with stink-bugs of various kinds, but none of them is recorded as having habits of any special interest. The European *Lindenius albilabris*, not uncommon in Britain, belongs to a tribe of wasps, the *Crabronini* to be further dealt with in the next chapter, mostly fly-catchers.[49] The *Lindenius* is one of those which departs from family custom in catching mainly stink-bugs. Yet it has not wholly departed from tradition, for it takes also a certain number of flies, which may be mixed with its bugs. The flies are of a family in which its nearest 'orthodox' relative, *Lindenius panzeri*, specializes.

In turning to the hunters of *Homoptera* we must give pride of place to the cicada-hunter, *Sphecius speciosus*. This is a large and very handsome American wasp, rusty-red down to the waist, its abdomen marked with black and yellow and its wings russet. There is a species of the genus in southern Europe also. *Sphecius* has bulky prey to transport. It hooks its middle legs under the cicada's wings, carrying it belly up. It may have to climb up a stem, or in one instance the observer's trouser-leg, in order to gain height for a successful take-off. As a rule two cicadas suffice to feed a female *Sphecius* larva, while males have to make do with one. Reinhard[18] describes how the young larva 'pours out solvent' so that it can 'drink' its victim: it does not really breach the cicada's armour till it is half-grown.

A number of wasps prey on the smaller Homoptera. The black and yellow *Gorytes* are conspicuous among these (Plate 6B). They burrow in sand and carry hoppers to their nests in flight, storing perhaps ten to fifteen to a cell and effecting a final closure of their nests by pounding with the end of the abdomen, as many pompilids do.[5] There are some European species which hunt for frog-hopper nymphs. One, *Gorytes mystaceus* (Fig. 5), is one of the commoner solitary wasps in Britain, being found in all sorts of country, not only where there are sandy commons. Frog-hoppers are familiar as adults with a rather squat frog-like form, rather less than half an inch long and with a remarkable power of jumping. The immature forms, the nymphs, are even more familiar as the producers of cuckoo-spit, frothy material found on various plants. It consists of the insects' watery waste-products whipped up into a froth, and inside it sits the soft green nymph. The froth

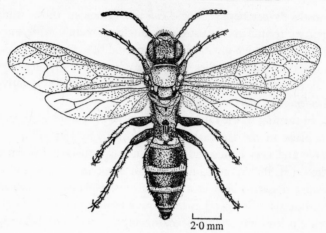

2·0 mm

5. *Gorytes mystaceus*

doubtless protects it from some enemies, but not from the *Gorytes*, which extracts it from the 'spit' and carries it off.[50] According to one account it pushes in its head and pulls out its prey: according to another it inserts its legs and sting to effect the capture.

Besides the inevitable shadow-flies which afflict so many wasps, the *Gorytes* suffers from the depredations of members of a closely related genus, *Nysson*, of which there are numerous species. In Europe, *Nysson dimidiatus* often seems to be commoner than its *Gorytes* host. It enters the *Gorytes* nest when the owner is not at home and lays an egg carefully concealed between the upper wings and the abdomen or other secret place and, for further concealment, on the bottom bug in the pile. If, however, the *Gorytes* is in the burrow, the *Nysson* will wait patiently outside a few inches away until the coast is clear. Other *Nysson* species go around raiding already closed cells. Females have been seen tapping the ground with the antenna and abdomen, and somehow detecting where the *Gorytes* burrow is. Then down they go, stay down perhaps half a minute to lay an egg, and out again, carefully closing the entrance when they leave.[5] Doubtless they feel that otherwise

some horrid parasite might get in! When the *Nysson* larva
hatches it makes its first meal off the *Gorytes* egg. One *Gorytes*
species in Germany is called the fetish-wasp from its fondness
for visiting the flowers of fly-orchids.[15] It is generally believed
that these orchids have their flowers in the shape of an insect,
with the intention that these will attract male insects on the
look-out for a member of the opposite sex.

There is a group of closely related wasp-genera, black or
black-and-red, slender, very wasp-waisted, rather like miniature *Ammophila*; these all hunt for small homoptera.[60] Two
species of *Psenulus* nest in burrows in wood or stems or in
straws and store large numbers of aphids, twenty-one to
thirty-six in one cell. Fabre watched one extracting aphids
from the galls made by the creatures in a turpentine-tree. The
wasps are very industrious and may continue working after
sundown. The very similar genus *Psen* seems sometimes to
nest in holes in wood and sometimes in sand. *Mimesa*, is a
genus consisting mainly of sand-burrowers, and many holes
may be found close together: a few species, however, nest in
wood. The wasps are often seen flying apparently aimlessly
back and forth over bushes. They hunt for leaf-hoppers. One
observer found that fourteen of nineteen hoppers of one species
which were captured by the wasp had already been attacked
by an internal parasite, the larva of another kind of wasp;
perhaps this had made their movements more sluggish so that
they were caught more easily. Some *Mimesa* are found
especially on maritime sand-dunes: one recently described
species is known only from dunes on the two sides of the
Irish sea.

There is little new to say about other genera of small black
aphid-hunting wasps, many of which are not uncommon.
Most of them nest in holes in wood or plant-stems, a few in
sand. The Peckhams[16] studied one, a *Passaloecus*. After excavating its burrows in wood, it took a whole day to rest; then
started bringing in aphids, one every four or five minutes and
then pellets of mud with which to close its nest.

Species mentioned

Astata unicolor	A	A-stat'-uh eu'-nick-o-lor
— occidentalis	A	— oxydent-ale'iss
— boops	B	— bo'-ops
Nysson dimidiatus	B	Niss'-on die-middy-ate'-us
Gorytes mystaceus	BA	Go-rite'-ees miss-tay'-cee-us
— laticinctus	B	— latty-sink'-tus
Sphecius speciosus	A	Sfee'-cee-us spee-cee-oh'-sus
Passaloecus	BA	Pass-a-leek'-uss
Psen	B	Pseen
Psenulus	B	Pseen'u-lus
Mimesa	B	My-mees'-uh

CHAPTER 10

Some Fly-catchers

THERE are so many and such various sorts of flies in the world that it is not surprising that many kinds of wasps prey upon them. Some provision their nests with almost any fly of about the right size, while others specialize in flies of particular families.

One of the commonest of solitary wasps in Britain and some other parts of Europe is *Mellinus arvensis*, a medium-sized insect, banded with black and yellow [50] (Plate 7A). It catches middle-sized flies of whatever kind happens to be plentiful near its nests, very often those of the Muscid family or flesh-flies (*Sarcophaga*), sometimes house-flies. It is fond of hunting near heaps of dung, sitting motionless close by and pouncing suddenly with its jaws wide open when it sees a likely victim. It lands on the fly's back, sometimes seizing a wing in its jaws and the body with its legs. Then the fly is manipulated into such a position that the sting can be applied under the thorax (Plate 7A). Stinging is repeated if necessary till the fly ceases to struggle. It is then turned over, the wasp's jaws grasp it by the proboscis and it is carried off in flight to the nest. Successful stinging seems to depend on coming into play of a series of reflexes. If the fly struggles partly free or gets into a position awkward for the wasp's performance, the latter abandons her original scheme, and kills and eats the fly instead. Malaxation may squeeze out some of the fly's body-juices, which are used as nourishment, and such a spoilt fly is usually discarded. One observer suggests that *Mellinus* destroys more flies by eating them than for provisioning her nest. Though the routine is to seize the fly by the proboscis and to grasp it for dragging into the nest, this is not invariable. When a fly at the nest entrance was snatched away and the proboscis cut off, the *Mellinus* dragged its prey in all the same.

The burrow is made by digging, using the head and fore-legs as a sort of tiny dredging-basket for sand. It is often on a steep or even vertical bank and if so the wasp lands with its fly at the bottom of the bank, dragging the prey up and turn-ing round as it enters the burrow. There are, according to most observers, cells branching off from a main tunnel; four to nine flies are placed, heads pointing inwards, in each cell. Very occasionally the wasp, probably inadvertently, lays two eggs in one cell, and it is likely that, in that case, one larva eats the other. The larva of *Mellinus arvensis* is unusual in having a purplish colour.

Among north American fly-catchers are fifteen species of the genus *Steniolia* and these show several points of interest.[39] They practise progressive provisioning, as do their relatives of the genus *Bembix*. They catch especially bee-flies (*Bomby-liidae*). These, as we have seen, are parasites of bees and wasps and it would seem a good plan to feed your young on the bodies of your enemies. There does not seem, however, to be evidence that the bee-flies in question are particularly apt to afflict the *Steniolia*. These Steniolias are unusual in having a proboscis much longer than that of most wasps and ap-parently particularly adapted to extract the nectar from various flowers of the family *Compositae*—thistles, daisies and so on. Most remarkable of all, however, is the habit of sleeping in large aggregations. There may be fifty, or as many as five hundred, in a cluster hanging from a branch of a tree (Plate 3B). In Wyoming the same branch of a lodgepole pine tree was occupied by a cluster of *Steniolia obliqua*, the com-monest member of the genus, during at least five consecutive summers. Other collections were on a sunflower's head which drooped down towards the ground under the weight of wasps. Both sexes are found in the cluster, the females tending to form tight balls in the middle with the males on the outside. This is perhaps because the males tend to come home to roost later than their women-folk. Matings occur not infrequently in and about the clusters and the only likely suggestion as to their meaning is that they are concerned with mating-behaviour – what might now-a-days be called a 'love-in'.

Wasps of the genus *Oxybelus* are small, dumpy but very active wasps. The European *O. uniglumis* is very common in sandy places: it has paired whitish spots on the abdomen.[50] The females commonly catch their flies, especially members of the family *Muscidae*, in the air. They may, however, 'run about on grass-stems and catch flies that come to bask, pouncing on them like a cat'. Others have been seen 'attacking flies settled on pine-stumps, catching them in the air, and also frequently taking those which were resting on the observer's hat, even when he was walking'. This attack is so rapid and successful that it has been difficult to see how the effective blow is struck. It seems, however, that the sting is inserted under the neck or thorax. The flies are killed or almost wholly paralysed. In transporting its prey, *Oxybelus* does something novel.[4, 32] The fly remains impaled on the wasp's sting and is carried like that, trailing behind the wasp in flight; the fly's head may almost touch the wasp's tail. When the nest is reached, the temporary closure is kicked away, the wasp using all its six legs and raising the fly in the air on the end of the sting. Four to twelve flies are stored in each cell. One observer saw the wasp bring in six flies in five minutes. The wasps sometimes steal flies from each other's nests, but if the rightful owner returns she may manage to retrieve her own property. At least two American species of *Oxybelus* are also known to transport their flies transfixed on the sting. A very silvery species, *O. argentatus*, is found on sand hills along the west coasts of Britain and France and apparently specializes in hunting a very silvery sand-dune fly, *Thereva annulata*.

Needless to say, *Oxybelus* is plagued by various parasites, especially the shadowing-flies. *Senotainia conica* follows *O. uniglumis* and is hard to shake off, and has been seen rushing into the burrow after the wasp in order to 'lay' its new-born larva. As many as three flies may hang around the entrance to a nest. (Plate 7B). They do not quarrel but take turns to go inside and achieve their fell purpose. A related fly, *Metopia leucocephala*, sometimes pounces on the prey as it is being dragged into the nest; at other times it follows the wasp in and allows itself to be shut in, so that it can operate and then make its way out at leisure.

The last fly-catchers we have to deal with belong to the members of the family *Crabroninae*. At one time nearly all of these were included in the genus *Crabro*, named from the Greek word for a hornet. They are, however, much smaller wasps than hornets and there must not be confusion between wasps of the genus *Crabro* and the true hornet *Vespa crabro*, to be considered in chapter 14. The big genus *Crabro* has now been divided into many smaller genera. Most of the smaller species included are black: most of the bigger ones are black and yellow. Most of them prey on flies, but, as mentioned in the last chapter, some species catch bugs; these are usually small hoppers. Some catch flies *and* hoppers.[49] One species (*Crossocerus walkeri*) preys on may-flies, another (*Lindenius armatus*) on parasitic hymenoptera such as ichneumons, another on moths. Among the more numerous fly-hunters there is a good deal of specialization, many of the species taking predominantly members of one fly family and even concentrating on one sex of victim. *Crabro scutellatus* goes for females of the *Dolichopodidae*, a family of rather long-legged metallic flies usually found near water. More than all this, there is a tendency for local races of a single species to take flies of one particular sort, and this seems to be not merely a question of making the most of a locally abundant source of victims. It has been suggested that wasps may prefer to catch flies with the familiar taste of the food that raised them in their larval days. One can see how such a tendency might lead in the course of evolution to specialization in prey-hunting.

The nests of some members of this family are dug in sand, while others use tunnels in wood or in plant-stems. In the latter cases there are likely to be cells in a single series. In the sand there is commonly a main tunnel with cells in side

branches. There may, however, be something more complicated. For several species it has been observed that a number of
females enter the same main burrow, and it is likely that there
is a primary main burrow, off which come secondary burrows,
the property of individual females, and off these again the
passages leading to individual cells. Evans[33] made a study of a
species with this behaviour, *Moniaecera asperata*, which burrows in clay-sand: and the habit is recorded also in Britain
for *Crabro leucostomoides* which nests in wood
and for *Crossocerus elongatulus* nesting in
crevices between paving stones. Adoption
of such habits may be one of the steps towards a truly social behaviour.

The Peckhams[16] watched a *Crabro sex-
maculatus*, digging its holes, spreading out
the excavated sawdust with its wings and
mandibles and then further dispersing by
fanning with its wings. They watched another species (*C. stirpicola*) excavating a
hollow stem. The burrows of this species
may be 30 to 40 cm long. The wasp they observed continued to dig for forty-two hours
on end: only one ten-minute pause was recorded. The amount of sawdust brought out
was measured by inverting a bottle over the
open top end of the stem and making secure

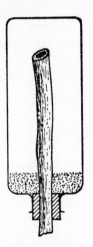

6. Experiment on
excavation by
Crabro stirpicola

the mouth of the bottle around the stem; the rate of accumu-
lation of sawdust could then be measured. (Fig. 6.) The little
wasp worked even faster during the night than in the day-
time. Alas, for all this hard labour: after it had provisioned
one cell, the wasp disappeared.

Species mentioned

Mellinus arvensis	B	Mell-ine'-us ar-venn'-sis	
Lindenius albilabris	B	Lin-dee'ny-us alby-lay'-briss	
— panzeri	B	— pann'-zer-eye	
— armatus	B	— arr-mate'-uss	
Crossocerus elongatulus	B	Kross-oh-seer'-us ee-long-ate'-u-lus	
— walkeri	B	— walk'-er-eye	
Crabro leucostomoides	B	Kray'-bro leu-kos-tom-oi'-dees	
— stirpicola	A	— stirpick'-oh-luh	
— scutellatus	B	— skew-tell-ate'-us	
— sexmaculatus	A	— sex-mack-u-late'-us	
Moniaecera asperata	A	Monny-ee-seer'-uh ass-per-ate'-uh	
Oxybelus uniglumis	B	Oxy-bee'-lus u-nee-glue'-miss	
— argentatus	B	— ar-jent-ate'-us	
Steniolia oblique	A	Stenny-oh'-lee-uh o-bly'-quuh	

Prey

Thereva annulata	Therr-ee'-vuh ann-yew-late'-uh
Dolichopodidae	Dolly-ko-poh'-did-ee

CHAPTER 11

Wandering Wasps

THE convenient but not very scientific term 'wandering wasps' was used by Iwata[51] to cover a number of relatively small wasp families. All of them are either parasitic or arrange for their descendants without the fixed homes which are the rule in the *Sphecidae*, which we have been dealing with in the last eight chapters.

Somewhat apart from the other wandering wasps are the Chrysids, ruby-wasps or ruby-tails, all of them parasites of other wasps or of bees. They are related to some small wasps, the *Bethylidae*, the larvae of which are internal parasites of caterpillars, as are those of ichneumons. Indeed the Bethylids look rather like ichneumon-flies and were formerly included among them. Very different in appearance are the Chrysids themselves, which are among our handsomest wasps. They are arrayed in glittering metallic armour of red, green and gold. The commonest species in northern Europe, including Britain, is *Chrysis ignita*. This has a brilliant blue-green head and thorax and an equally brilliant metallic red abdomen. It may often be seen flying up and down walls or gate-posts looking for the holes of its victims. I have deliberately used the word 'armour', for these wasps have extremely hard armour-plated skins, as is known by any collector who has tried to insert a pin through the thorax. They need their armour, for their way of life involves the entering of the nests of wasps and bees, all armed with stings and ready to attack an intruder. A number of chrysids have abdomens which are concave beneath; if the wasp is confronted with an attacker it can bend its head and thorax forward into the abdominal concavity, thus rolling itself up into an invincible armour-plated ball.

Most of the chrysids are parasites of particular species of wasps or bees or perhaps of a few closely-related species. Some, such as *Chrysis ignita*, have both bees and wasps as their victims. It would be very odd if such wasps behaved like the cuckoo-wasps we have considered earlier which quickly destroy their host's egg or young larva and consume its food-store. This would mean that the chrysid larva would some-times be raised on paralysed spiders and flies and sometimes on stored honey: an improbable state of affairs. These chrysids on the contrary are not cuckoos, but rather parasites or parasitoids, for they feed, not on food-stores, but on bee or wasp larvae, often fully-fed ones. There are, however, other chrysids with a cuckoo-like behaviour.

Ferton has described the habits of some species of *Chrysis* parasitic in France on bees of the genus *Osmia* (see chapter 19). *Chrysis dichroa* (two-coloured) attacks the red-haired *Osmia rufohirta* in Corsica.[35] This bee makes its nests in empty snail-shells. The Chrysis has a long telescopic ovipositor and lays its egg when the *Osmia* is away from home, placing it at the opposite end of the honey-mass from the bee-egg. When the bee-larva hatches, it has to eat its way right through the honey-mass before it encounters the chrysid larva. So the latter then finds a nearly full-fed larva ready to be attacked. It not infrequently happens that a cell receives the eggs of several *Chrsyis*: as many as six in a cell have been recorded; yet never does more than one reach maturity. The first task of a newly-hatched chrysid larva is to see that it has the field to itself; competitors must be eliminated. Young chrysid larvae are very different in appearance from those of other wasps. They are covered with hairs, which perhaps protect them from being bitten, and have useful antennae: they also have sharp sickle-shaped jaws and they are fairly mobile. If one chrysid larva meets another, there is a fight and only one survives. Perhaps the odds are on the one which happens to hatch first. Nineteen days or so after hatching, the larva's moult turns it into something very much like any other wasp

5(A) Bee-wolf, *Philanthus triangulum*, taking a honey-bee into its burrow
5(B) *Cerceris rybyensis* with its *Halictus* victim

6(A) The bug-hunter, *Astata boops*

6(B) *Gorytes laticinctus*, huntress of cuckoo-spit nymphs

larva. The surviving larva proceeds to consume all of the bee larva leaving nothing but empty skin.

Another species of *Osmia* (*O. saundersi*) burrows in hard sand and lines its nest with petals of a yellow sun-rose (*Cistus*). Its enemy is *Chrysis prodita*.[46, 47] This wasp burrows down, seeking the bee's cells. It then pierces the cell and breaks into the cocoon through a weak spot which has been left to further the ventilation of the interior. Eggs are laid only on a bee larva which has finished feeding and has spun its cocoon. If that stage has not been reached, no egg is laid and the wasps explore elsewhere. The job is done very meticulously; a wasp has been timed, spending more than an hour underground. The newly-hatched *Chrysis* larva spends a little time seeking out and destroying any other *Chrysis* eggs before getting down to eating its bee-larva; this it finishes off in six days.

Across the Atlantic Dr. Krombein[13] found out much about chrysids in his trap-nest. (p. 45). One species, the dark-winged *Chrysis fuscipennis*, breaks into the cells of mud-daubers. Those haunting the trap-nests behave otherwise. The blue *Chrysis coerulans* waits till its victim has finished its provisioning. This is the Eumenid wasp, *Ancistrocerus antilope*, to be described in the next chapter (p. 92). The chrysid chooses the best moment by waiting till the *A. antilope* has begun to bring pellets of mud for closing the nest. The host wasp has several generations in a year, and so has its parasite; the development of host and parasite run parallel, so that everything is nicely synchronized for the chrysid's benefit. It may, however, happen that when the chrysid wasps emerge from their cocoons, they find the way blocked by wasp-larvae in other cells which have not been parasitized; then, on their way out, they eat enough of the obstructing wasp-larvae to give a passage-way.

Chrysura kyrae was found as a parasite of *Osmia lignaria* in Maryland. This bee made its cells in the holes of the trap-nest in a linear series, as, very naturally, it had to do. The *Chrysura* kept the nest under surveillance, popping in as opportunity

6

arose and managing to lay in many cells. In one nest with eleven cells, there was a *Chrysura* in cells five, eight, nine and ten. In another with twenty-three cells, numbers seven, thirteen, fourteen, sixteen and twenty-three all had a *Chrysura* in them. If the bee discovers that it is being imposed upon, it will sometimes stop building cells in that nest, close it prematurely and proceed elsewhere. The eggs are laid on the side of the pollen mass, so that the bee cannot see them. About a week after it hatches, the *Chrysura* larva finds its victim and attaches itself by its mandibles to its back or side. When the bee-larva moults, it moults too and afterwards quickly manages to re-attach itself. When the bee-larva spins its cocoon, the chrysid rapidly polishes it off and spins its cocoon inside the bee's cocoon.

Other chrysids take little or no food from the growing host larva, but wait for their feast till the host larva is full grown and spinning up. Krombein also describes the habits of some of the chrysid larvae which are cuckoos rather than parasites. One species sucks the egg of a *Trypargilum* (p. 46) and then consumes its store of spiders. Others deal similarly with the caterpillar-stores of *Ancistrocerus* and other Eumenid wasps. Yet these beautiful but ruthless wasps do not have it all their own way; their larvae may in turn be destroyed by the ubiquitous miltogrammine flies.

Wasps of a family closely related to *Chrysis* are called *Cleptes*, thieves. (Fig. 7.) They also are parasites of other hymenoptera, but this time of saw-flies. They discover, break open and lay their eggs in the cocoons of saw-flies of the genus *Nematus*, particularly those of the gooseberry saw-fly, *N. ribesii.*

The remaining wasps to be considered in this chapter belong to a number of families all contained in one large superfamily, the *Scolioidea*, named after its characteristic genus *Scolia.*

Scolia species are hunters of beetle-larvae, particularly subterranean ones such as cockchafer-grubs. They are large,

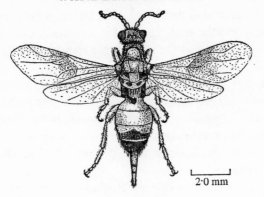

2·0 mm

7. Cleptes semiaurata

stout, often hairy wasps, mostly black with some red or yellow markings. The female burrows down in August and September, finds, stings and paralyses her prey and lays her eggs. The wasp evidently spends much of her time underground in the 'hunting season'. Trails up to twenty inches long have been traced in the earth where she has passed. As was described for some pompilids, there is no attempt to carry the larva off to a nest. The wasp larva is left to consume its beetle larva in *situ*, even though this may be smothered in earth. The stinging is no easy matter; Fabre[11] records that the wasp may have to struggle with her victim for a quarter of an hour before finding the right spot to insert the sting between the joints on the underside of the thorax. He also records that when one species was offered a beetle-grub which was laid down on a smooth surface, the larva remained curled into a ball, hedgehog-like, and the wasp could not reach the vulnerable spot. When, however, the larva was placed on sand it uncurled and the wasp readily stung it.

The egg is laid on the beetle-grub's belly. A rose-chafer larva may be six or seven hundred times the bulk of a newly-hatched *Scolia* larva feeding on it. When the larva hatches, its anterior part is thin and snake-like; behind it is stouter. The head enters the host's body and is not withdrawn again. Fabre suggests that it takes first blood, then fat and muscle and only at the very end the vital organs. If it is moved it can't be successfully replaced and the interference may involve the death of the victim. Once that is dead, the *Scolia* larva dies also, for it must have fresh food. In order to spin up, the full-fed Scolia first makes a sort of scaffolding in the earth, then slips itself into a hammock and weaves a cocoon of two layers, the inner one being varnished. At one end is a circle of weaker resistance, so that a cap can be pushed off when the wasp wants to emerge.

It may be mentioned that one *Scolia* species was introduced from the Philippines to Hawaii to control a beetle which was a pest of sugar-canes there.[18] Such attempts at biological control sometimes recoil and turn out to have unexpected and disastrous consequences; but this particular attempt was successful.

Slender black wasps of the genus *Tiphia* also provide beetle larvae for their offspring. Of two British species the larger is coastal in its habits; a smaller one is found also inland. Both are fond of visiting the flower of *umbelliferae*, especially wild carrot or Queen Anne's Lace (*Daucus*). Their habits are similar to those of *Scolia*, but with some species the beetle-larva is not killed but carries on feeding while the *Tiphia*-larva feeds on it. The smaller species, *T. minuta*, may prey on larva of dung-beetles (*Aphodius*).

The wings of these wasps must be rather inconvenient if there is much underground-hunting to be done, and so we find for a number of species that the males are winged and the females wingless. Such a wasp is the European *Methoca ichneumonoides*, ichneumon-like as the name implies; it is a hunter of the larvae of tiger-beetles, *Cicindela*. These are predaceous both in the larval and adult state. The larvae live in burrows, sometimes a foot deep, in sand. They fix themselves at the mouth of the burrow, blocking the entrance with the head and thorax and wait for some unwary insect to come within range; it is then quickly snapped up. The *Methoca* female, black and red and slender, wanders around till she finds a tiger-beetle's burrow. She slips into this and attempts to sting the larva under the neck. Sometimes she does this without any trouble. At other times the tiger may succeed in seizing her. This does not necessarily bother her as she can then curve round her abdomen and administer her sting. Again, there may be a long rough-and-tumble before the tiger is paralysed. After stinging its victim the *Methoca* retreats to the top of the burrow and waits for ten minutes or so before cautiously descending to see whether the sting has been effective. If all is well she squeezes past the larva, administers more stings for good measure and malaxates the region between its legs. The egg is laid, *Methoca* withdraws and carefully closes the burrow. It may be asked how we know about all these subterranean dramas. It is through the patient observations of several observers who have kept the *Cicindela* larvae in sand in glass jars so that one side of the burrow is against the glass and so available for inspection. One of these observers, H. T. Pagden,[54] records that the *Methoca* occasionally passes the night in a tiger's burrow without first attacking it. Perhaps the beetle larvae were of the wrong size. Pagden adds: 'why they did not attack the *Methoca* I am unable to say'.

A number of entomologists have noted that on sandy commons in Britain *Methoca* females are not uncommon, yet the winged males are very rarely encountered. It has been shown by breeding experiments that the species may be parthenogenetic, that is the unmated female can lay fertile eggs. In this case and in contrast to what is more usual among hymenoptera, the unfertilized eggs usually give rise to females.

The next group is that of the *Mutillidae* or velvet-ants. All have wingless females. In Britain there are two species, one twice the length of the other. The big one, *Mutilla europaea*, has a reddish-brown thorax and a black abdomen of velvety texture, with white markings. In shape it looks much like an ant: hence the name 'velvet-ant'. The female is a parasite of bumble-bees, particularly the reddish-brown species

Bombus agrorum and *B. humilis* (see chapter 21). It is a local insect, but is very occasionally so successful that a *Bombus* nest may yield more velvet-ants than bumble-bees.

Mutillids are much commoner in countries warmer than Britain, and may be very troublesome to other hymenoptera on the European continent and in the United States. Like the Chrysids they have hard outer skins, not readily pierced by a hostile sting. Some are host-specific but others are catholic in their tastes and some may parasitize either wasps or bees. The genera *Crabro*, *Oxybelus* and *Bembix* are among those attacked. As with the chrysids, the female mutillids lay their eggs on full-grown resting larvae. They may be seen crawling around on the ground looking for nests. They may then burrow down, breach the cell wall and cocoon to lay their egg. Krombein[13] studied some species which entered his trap-nests. He noted that they were not very efficient in that they only attacked the outermost cell in a linear series. After opening a cell and the cocoon within it and laying their eggs some mutillids then sealed up breaches in the cocoon, but failed to do the same for partitions between cells and for terminal plugs; so there was little to prevent other parasites from getting in later. A female mutillid has been seen, before laying

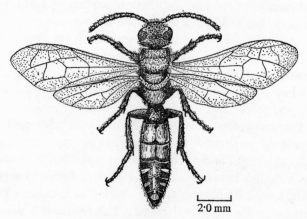

2·0 mm

8. *Sapyga quinquepunctata*

her own egg, making a thorough inspection apparently to make sure that no other parasites had got in before her.

Mutillids are among the wasps which may destroy more of their victims by killing and eating them than by storing them for their larvae. Evans[5] even accuses one of stealing into a *Bembix* burrow at night and sucking at the owner's throat: a veritable vampire.

The last of our families of wandering wasps is a small one: the *Sapygidae*. *Sapyga* species (Fig. 8) are black and yellow or black, white and red wasps and are cuckoos in the nests of *Osmia* and other long-tongued bees. Both sexes are winged. When the cuckoo's egg hatches, the bee's egg is first destroyed and then the *Sapyga* larva proceeds to consume the pollen-nectar mixture which the bee has stored.[13] This it does rather slowly, perhaps taking three weeks over its meal. We thus have a wasp which, in contrast to all those so far considered in this book, is reared on vegetable food like bees and not on insects or spiders.

Species mentioned

Scolia	E	Scoh'-lee-uh
Tiphia minuta	B	Tiff'-ee-uh mi-nu'-tuh
Mutilla europaea	B	Mu-till'-uh
Methoca ichneumonoides	B	Mee-thoe'-kuh ik-nu-mon-oy'-deez
Sapyga	B	Sapp-eye'-juh
Chrysis ignita	B	Cri'-sis ig-nite'-uh
— dichroa	E	— die-crow'-uh
— prodita	E	— prodite'-uh
— fuscipennis	A	— fuss-ki-penn'-iss
— caerulans	A	— seer'-u-lans
Chrysura kyrae	A	Cri-syure'-uh ky-'ree
Cleptes	B	Klep'-tees

Species mentioned (*cont.*)

Prey

Aphodius   Aff-oh-'dee-us
Cicindela   Sy-sin'-dell-uh

For pronunciation of other host species see lists at end of chapters 12, 19 and 21.

CHAPTER 12

The Furies: Potters and Masons

THE ancient Greeks believed in some ferocious females, known to us as the Furies; only the Greeks were not rash enough to call them that—they hoped to placate them by calling them Eumenes, well-disposed ones. The wasp-genus *Eumenes* is the type of an important family, *Eumeninae*. These wasps are of much interest, but not because they are either specially furious or particularly well-disposed. They belong to the larger family of *Vespidae* which includes the social wasps: all these, unlike the wasps considered in earlier chapters, pack up their wings, when at rest, in longitudinal folds. The Eumenids, being like the social wasps in structure, are almost certainly on the branch of the family tree which led to the evolution of those highly successful insects. In their habits, however, they closely resemble the common run of solitary wasps. A very few have broken away from orthodox behaviour.[14] An African *Odynerus* (Fig. 9) practises

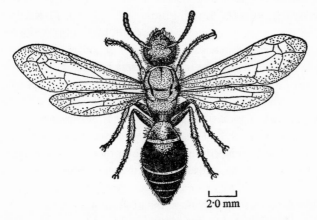

2·0 mm

9. *Odynerus spinipes*

progressive provisioning and a *Stenogaster*, an old world tropical species, cares for several nests at once: both steps in the right direction. A *Synagris* species goes definitely farther, feeding her offspring daily on chewed-up caterpillars and some *Stenogaster* species feed their larvae chewed-up midges, which they snatch out of spider-webs.

The European and North American species are, however, in this respect old fashioned and orthodox. In construction of their nests, however, some of them show remarkable orginality. Wasps of the genus *Eumenes* itself differ from others in the family in having a very narrow elongated first segment of the abdomen. They, and in particular the European *E. coarctata*, are the real potters.[6] This species is found in heathery sandy places and she builds her nests of mud, fixing it to heather-stems or other suitable bases of attachment. (Plate 8A.) Pellets of mud are brought in the mouth and built into a hollow sphere, and then at the mouth of the sphere is an expanded neck, so that the whole looks like a rather dumpy flask or amphora. Each cell accommodates one larva; there may be several closely adjacent cells. The neck of the flask is so narrow that the wasp herself cannot get in. There is room enough, however, for her to insert the little caterpillars needed for the larva's food and also to introduce her ovipositor and lay an egg. Besides bringing mud for building, *Eumenes* and other potter wasps convey water as required to bring the mud to the right consistency. The cells are of course closed with a door of mud. All of the mud duly sets to form a hard shelter for the growing larva. It may be strengthened by incorporating bits of quartz, flint or snail-shells. Fabre credits a French species with some artistic sense! There is, however one member of the family, *Odynerus vespiformis*, which habitually breaks down finished mud-pots of other species, bringing water to soften the walls and thus making a hole from which the contents can be extracted.

Other Eumenids burrow in the ground or use holes in wood or stems. Some of the former usually make nests on steep banks but sometimes on flat ground, and construct elaborate tubular entrances of mud. These may take the form of turrets, which may be vertical at first and then turn down. (Plate 8B.) A tube may be several inches long. Olberg[15] speculates as to the object of these structures. Is it to help the wasp to find its home, to thwart parasites, or as a convenient dump for surplus building material? Whatever it is, the wasps have been seen, after provisioning perhaps several cells in the burrow beneath, dismantling the tube erected with such labour. The material obtained from its demolition is used in closing the burrow.

Yet others in the family make use of any cranny or hole and use their masonic talents to turn it into a nest. Some middle eastern species used the furrows of hieroglyphic writings on stones, filling them up with clay for their homes and obliterating them.[18]

The commonest prey in this family consists of caterpillars, especially those of geometers or loopers. Before, however, these are hunted, the egg is laid and in a very characteristic manner. It is suspended by a delicate filament from the roof or sometimes the side of the cell. When the larva hatches, it does so almost imperceptibly, not completely leaving the egg-shell. It reaches out to its first caterpillar and begins to take a meal. As this progresses, the larva comes farther and farther out, and the egg-sheath forms a sort of ribbon which further extends the region in which the larva can operate. Fabre[6] describes how it slides up and down this structure, withdrawing when it has for the time finished eating.

The caterpillars which are stored in the nest are often incompletely paralysed and, if the cell has been imperfectly sealed, may even escape. Fabre supposed that the object of the egg's suspensory thread was to preserve the young larva from possible squashing by wriggling caterpillars. He tried in vain to rear *Eumenes* larvae with methods successful with other wasps and ascribed his failure to the fact that the thread was missing in his experiments. Other workers have been reluctant to accept Fabre's conclusions: eggs have been separated from their threads and placed among the caterpillars; the wasp larva seems not to have been inconvenienced.

To return to the wasps out on the hunt: the caterpillars brought home are usually small ones and as many as twenty may be stored in one cell. Some species go for leaf-mining larvae, cutting open their mines and stinging them, usually several times.[18] Larvae may be bruised by malaxation. They are carried in, belly to belly. Where larger larvae are being transported, flights may have to be in several stages.

When adult wasps finally emerge from their cocoons they commonly rest for a day or two to get hardened up. If the cells are in linear series, the males are usually in those nearer the entrance and come out first. Where there are underground burrows, gravity tells the emerging wasps which way to go to reach the outside world. But what about the larvae in horizontal burrows? If they tried to make their way out in the wrong direction they would come up against a blank wall. It is found that if trap-nests are examined, the pupae are almost always facing the right way. K. W. Cooper[28] examined 2,674 cells, mostly of *Ancistrocerus antilope*: only fourteen out of all these were facing the wrong way. Now the larva when making its cocoon twists about in all sorts of ways; it could not possibly keep a sense of direction from the time it hatched from the egg. The explanation is that the mother wasp in building her cell leaves a message for her offspring, telling it which way to get out. This is done by fashioning the two ends of the cell rather differently. The end to which the pupa must face and through which the perfect insect must emerge is made concave and also smoother than the other end. It was shown by varying the nature of the ends that the 'concavity' signal was more important than the 'smoothness' signal. All this was revealed by experiments in which the direction of cells in the series was reversed. If the wasp pupae were turned round in their cells, only 50 % succeeded in emerging. An interesting point is that quite unrelated wasps seem to understand the same 'language', which K. W. Cooper calls 'digital communication': *Trypoxylon* larvae (chapter 6), when they spin up, know in just the same way in which direction to face. So does a chrysid parasite *Tetrachrysis*, but for once in a way the miltogrammine flies are caught napping and many die without finding the right exit.

Besides these flies and the ruby wasps which afflict so many hymenoptera, the Eumenid wasps are subject to the attentions of another sort of parasite, not hitherto mentioned,—mites.[27] Some of these have evolved a very special and remarkable relationship with their wasp hosts. Some of them suck blood from the resting larva or pupa without actually killing it. Others devour both the food stored for the wasp-larva and the larva itself. Others are apparently only scavengers, cleaning up debris. Mites are also found on the adult wasp and some wasps have actually evolved a special cavity on the abdomen for sheltering mites. One of the basal abdominal segments overlaps the one below, making a convenient hollow for this purpose. One of the larger Eumenine wasps is *Monobia quadridens*: it is marked with black and white instead of the more usual black and yellow. The immature mites feed on debris in the cell and collect in the genital chambers of the male wasps when these hatch out. They then make their way on to the female wasp when mating takes place.[13] Females of *A. antilope* when they hatch out of the cocoon go to the trouble of killing and eating all the mites in their cells before they emerge into the outer world. What good this does is obscure, for they all acquire numerous other mites from the male wasps when they mate. Such mites soon reach the special mite-sanctuary on the wasp's abdomen, where they sit—maybe a hundred or more—in 'shingled ranks'. One of these mites has a remarkable life history. The male mite develops within the body of the female. When he comes out he mates with his mother, who then proceeds to lay fertile eggs on a newly transformed wasp-pupa.

Species mentioned

Eumenes coarctata	B	U'-men-ees co-ark-tay'-tuh
Oplomerus reniformis	B	Op-lo-meer'-us
Stenogaster	Tropics	Stenn-o-gass'-tuh
Synagris	Tropics	Sinn-agg'-ris
Odynerus vespiformis	A	O-dinn-eer'-us vesp-ee-form'-iss
Ancistrocerus antilope	BA	An-siss-tro-seer'-us an-till-oh'-pee
Monobia quadridens	A	Mon-oh'-bee-uh quod'-ree-dens

Parasite

Tetrachrysis	A	Tett-ra-cri'-sis

CHAPTER 13

Paper-wasps

THE genus *Polistes*, (Plate 10A) comprising the paper-wasps, is so called because of the paper-like material of which they build their nests. The social *Vespa* and *Vespula* wasps to be considered in the next chapter use similar material, which is made by chewing up wood into a pulp and letting it set. *Polistes* is a genus of special interest, for it affords clues as to the development of the most advanced social habits in the course of evolution. These wasps are largely inhabitants of warmer regions. There are many species in the tropics, and some in North America and Europe. The European *Polistes gallicus* is of only occasional accidental occurrence in Britain.

Female *Polistes* having been fertilized in the autumn, go into hibernation during the cold months, very often, with some species, in houses, They are then all ready to begin breeding in the spring. They are easy to observe, for the common *P. fuscatus* in North America and *P. gallicus* in Europe are very fond of building their nests in the roofs of porches, where they are conveniently sheltered from the weather. Nests may be started by single females or several may co-operate.[14] Such collaborative behaviour is commoner in tropical and subtropical species. There is commonly one laying female of *P. gallicus* in a nest in northern Europe, but several in the south.[20]

The nest begins as a paper stem, usually pointing downwards, and a cell, at first like a shallow cup, is suspended from this: further cells are added at the side until a regular comb is formed. A comb with 1,575 cells is recorded for one species. The egg is laid, and when it hatches, the young larva is provided with bits of catapillar and other insect food, the provisioning being of the progressive kind. As the larva grows, the height of the cell walls is increased by addition of more paper. The larvae hang upside-down in the cells, being at first anchored at the tail end with a sort of glue. Later they bulge out and fill the cell and there is no danger of their falling out. When the larvae are fed, the wasp bringing the food tickles or touches the larva, which secretes a drop of fluid from the mouth, and this the adult wasp consumes. There is a good deal of interchange of food between adults and larvae. Some partly chewed food given to a larva may after a little while be removed and passed on to another larva and so on.

The eggs laid early in the year all yield females. These look very much like their mother, yet their ovaries are imperfectly developed, they never mate and they are in fact workers. This differentiation is a big forward step on the road to the most advanced social life. In contrast to *Vespa* and the social bumble-bees and honey-bees there is very little difference in size between the queens and the workers of *Polistes*.

In the instances of collaboration between several laying females or queens there may be a hundred worker cells built before any young wasps hatch out. When these do hatch they complete any cells which the queen has not finished and undertake the collection of food and feeding the young larvae. Gradually the queen ceases to undertake these tasks, devoting all her energies to egg-laying. As the wasps hatch out, more cells are built to enlarge the comb, but also the cells vacated by some newly-hatched wasps may be used over again for another 'crop'.

Soon after mid-summer fertile females and males begin to hatch from a number of the cells and in due course mating takes place. The nests disappear, the males and workers die off and the mated queens go into hibernation. They may do this singly or in groups, in houses, under the bark of trees and elsewhere. Queens of several species of *Polistes* have been found hibernating together.

Though the workers cannot mate, they can lay eggs and these, as is usual with unfertilized eggs, yield only males. Occasionally a *Polistes* nest is found in which only males are being

7(A) *Mellinus arvensis* stinging a fly

7(B) The fly *Senotainia* on the watch for a victim

8(A) Potter wasp *Eumenes pedunculatus* carrying a looper-caterpillar into its pot

8(B) *Oplomerus reniformis* on its turret

produced. It is presumed that the queen has died or disappeared and that only workers are laying. So long as the queen
is in full activity she does her best to stop the workers from
laying. They have to enter a cell backwards in order to lay;
so she can catch them in the act.

The attachment of the queen to the nest is at times inconstant—perhaps an indication that a social existence is in the
early stages of evolution. Where there is some co-operation
between queens, one of them is dominant and does most of
the egg-laying.[4] She seems to be the one with the most fully
developed ovaries and she is able to overawe the others.
Below her there is a second in rank and so on—in fact a sort
of peck-order. A queen in the spring may start a nest of her
own and if joined by others becomes presumably the No. 1
queen. If others do not join her she may abandon the nest she
is starting and join another group. It is thought that queens
which come best through the winter with still some reserve
of fat are likely to be the ones to come out on top. Sometimes,
for some unknown reason, a top queen will leave her nest,
and No. 2 will then step into her place. It is recorded that
with *P. gallicus* in northern Italy, the top queen may achieve
such ascendancy over her rivals that these revert to menial
tasks of foraging; their ovaries may even undergo some de-
generation so that they virtually become workers.

Richards[20] from study of the behaviour of different *Polistes*
species, or even the same species in tropical and temperate
climates, has formed the view that social life in hymenoptera
may well have evolved in the tropics. In more southern countries
the hazards are more from enemies than from the weather. So
laying queens are produced earlier and there are relatively
fewer workers. In the Sahara new colonies are formed by
the swarming of young females along with workers, much as
will be described later in chapter 24 on honey-bees. In this
way, an insurance against destruction of a colony by enemies is
made more possible because of the early production of queens.
Farther north, on the other hand, rearing of a new brood is

7

more hazardous because bad weather may interfere; it is therefore better to have more workers so that growing larvae can have better care. There is a clue as to how this difference is brought about. If the wasps were kept warmer at night, at 25°C, the grubs were more apt to develop into queens; a lower temperature (5°C) led to more production of workers. In the tropics, nesting may go on throughout the year, new nests being started as a result of swarming and old ones not necessarily coming to an end at a particular time of year.

As might be expected with a large and successful genus like *Polistes*, there are species which have deviated from standard behaviour. One species in America (*P. perplexus*) is thought to be a cuckoo at the expense of other *Polistes* and three species in Europe have similar habits. Then there are a few which fill their cells with honey for the benefit of their larvae.

Mention must be made of some observations of the Raus[17] on the powers of navigation of *P. pallipes*. It is clear that, as has been described for other wasps, a locality has to be learnt. Young workers failed to find their way home when caught and liberated only an eighth of a mile away: one could not even manage 50 feet. Older workers had no trouble over distances such as these, but queens were generally more adept than workers. In one experiment 24 of 33 queens got home from the point of liberation while only 28 of 112 workers found their way from the same point. Queens were taken in darkness along a twisting route and got back to the nest from an eighth of a mile away in 22 to 72 minutes. When let loose they flew off in various directions but evidently soon recognized land-marks they knew. When taken a trip of 2½ miles, they were much less successful: some took 2½ days and others never returned at all. Males which were moved from their nest never managed to get home. Fabre and others have thought that this navigation was accomplished through some unknown sense. If such existed it is not resident in the antennae for wasps with antennae amputated could still home successfully. It is now generally agreed that there is nothing very mysterious about it all; *Polistes* like *Philanthus* and other wasps, is good at memorizing features of the landscape round her home.

Species mentioned

Polistes gallicus	E	Poh-list'-ees gall-'ick-us
— fuscatus	A	— fuss-kate'-us
— perplexus	A	— per-plex'-us
— pallipes	A	— pally'-pees
— nimpha	E	— nim'fuh

CHAPTER 14

Social Wasps or Yellow-jackets and Hornets

WE NOW reach the most familiar, most truly social wasps, those at the top of the wasps' evolutionary tree, the genera *Vespa* and *Vespula*. The English names of these insects are not quite the same on the two sides of the Atlantic. To most Britons the word 'wasp' means a social wasp of the genus *Vespula*, one of those common insects which get into the jam or make themselves a nuisance at picnics. Again, to a Briton, a hornet is the rather larger brown-and-yellow insect known as *Vespa crabro*. In North America most of the Englishman's 'wasps' are called 'yellow-jackets'; the term 'hornet' is usually applied not only to *Vespa crabro*, which is an introduced species in America, but also to certain rather larger *Vespula* species which are black-and-white in colour instead of black-and-yellow. The *Vespula* genus some time has separated off from it, as *Dolichovespula*, those wasps which, though similar in appearance to the others have different nest building habits: instead of building underground, they suspend their nests in bushes and trees. In this chapter the word wasp will be used for *Vespula* species and not in the more general sense as in previous chapters.

As with the paper-wasps, *Polistes*, females of *Vespa* and *Vespula* are fertilized in the autumn, and these fertile queens are the only wasps to survive the winter. Colonies persist through the winter only in hot countries. It is hardly desirable to speak of hibernation, for with some species males and females are produced before the end of the summer and the fertilized queens go into hiding and become dormant while the weather is still warm. More often, however, it is the chills of autumn which make them seek a suitable shelter. These queens not infrequently enter houses and may be found in a lethargic state in a fold of a curtain or elsewhere: if not wholly dormant they may give a sting at a time of year when this is not expected. Outside, the resting places chosen are often wholly unsuitable and many queens die before the winter is over.

When spring arrives they come forth and may sit and sun themselves when it is warm enough, but they do not immediately start to nest. In April and May they do begin to search for nesting places, most species seeking underground cavities. The hornet prefers hollow trees. The site chosen is often far from ideal, perhaps too exposed, but once it has been selected, a queen does not readily move. The underground hollow must have a root or a stone in the roof from which the nest may be suspended. It is made, as is that of *Polistes,* of 'paper' made of chewing wood and incorporating saliva to make a quickly drying pulp. Of the two commonest British species, *Vespula vulgaris* chews up rotten wood to make a yellowish pulp; *V. germanica* goes for sound wood and turns it into a tougher, greyer, pulp. The stalk from which the whole nest is to be suspended is at first flat and blade-like, then it narrows to a cord and finally expands again to make the first cell, which is long and narrow. Other cells, like crescents, are added to the sides of the first, and initially ten to twelve are built, sometimes more (Fig. 10). Soon an envelope or canopy is built

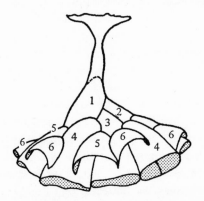

10. Early stages of *Vespula* nest

around the group of cells. The figure has been modified from one in a full account by C. D. Duncan[3] of the habits of *Vespula pennyslvanica* in California.

Next come the tasks of egg-laying and provisioning. The first cells will yield only infertile females or workers. To rear these the queen has, of course, to work alone. When they hatch out, they will do the outdoor work and she can concentrate on laying eggs. But this early period is very critical. The work has to be done at a time of year when the weather is cooler and often uncertain, and at this stage many nests come to grief. Everyone knows that there are good and bad wasp years—good or bad according to whether they are regarded from the point of view of man or wasp. In some years wasps are so numerous as to make outdoor picnics far from pleasant. It has been found that whether there are very many wasps in late summer, or relatively few, depends upon the weather in late April and May. Wasps are in particularly great trouble if an early spring stirs them up to make an early start and there follows a prolonged inclement spell. So one can often prophesy from seeing the early summer weather that there will be a good year for the wasp or a bad one.

The wasp larvae hang downwards in their cells, being anchored by thick mucus around their tails. While adult wasps feed only on liquid food, particularly nectar, the larvae take solids. This food consists of chewed-up insects, bits of meat with perhaps some fruit or regurgitated honey. In fact the diet is far more varied than for any of the solitary wasps. The insects taken are flies of all families and all sizes, caterpillars, less often beetles or bugs. When workers are feeding the larvae, they tickle or stroke them and stimulate them to secrete from their salivary glands a fluid which the workers consume. It is suggested that there is thus an exchange of food between worker and adult, a process referred to as trophallaxis. It may be, however, that this is only a way by which the larvae dispose of surplus fluid. Larvae are said to scratch the papery walls of their cells, making a rustling sound to attract feeders to them.

Hunting habits are rather different from those of solitary wasps. Species of *Vespula* hunt on the wing, flying rather high and pouncing suddenly on their prey, which they recognize by rather generalized features of colour and form. Sight plays a major role, and scent only when they get quite close. The prey is not stung but is killed by biting and then carried to a suitable resting place.[14] Some fluid may be imbibed from it and it may be cut up and chewed. Wings of a fly are often removed; they are obviously of no nutritional value. If the victim is too heavy to carry away, some is left to be retrieved on a later trip: the wasp then has to make a locality study before returning to the nest with the first load.

The first group of cells forms a miniature comb. As soon as the workers hatch out, things begin to move quickly, and as more and more cells are added, the comb grows fast and the nest has to be enlarged. When a sufficient number of cells have been added to the first comb, another storey is built below, suspended from the first by another stalk; and in due course a third or more may be added. There may be lateral supporting structures. More and more layers of covering paper canopy are added (Plate 9B). The cavity is enlarged by removing earth or other material, carrying it and dropping it at some distance from the nest. Hornets in their hollow trees gnaw away wood to enlarge their living space. It also becomes necessary to make room by removing the inner layers of canopy; and material so obtained is re-chewed and used to construct further canopy-layers outside. In a nest of the black-and-white American 'hornet', *Dolichovespula arenaria*, there had apparently been at one time no less than fifty-five canopy layers, of which thirty-five to forty had been torn down in this way. The canopies come to surround the combs completely except for the entrance, which is tidily finished off (Plate 9A). Sometimes there are two entrances. The tree-nesting species may have a tubular entrance, but it is not so neatly finished as with the underground nesters; the entrance may be at the bottom to start with, becoming lateral later.

There is no rigid division of duties among the workers, though probably the younger ones tend to receive the food and pass it on to the larvae and the older ones tend to do the foraging. At times a few will act as guard and, when it is very hot, some will engage in fanning to cool the nest down.

The queen deposits eggs in the cells with great care, glueing them to the side of the cell-wall. She may lay as many as 25,000 in a season. There is, of course, great wastage of workers and it is rare for there to be more than 5,000 present at any one time. The newly hatched larva is hard to distinguish from the egg: it moults three times before becoming a pupa. The larval stage lasts for twelve to eighteen days in the case of those destined to become queens, considerably less for workers. The pupal stage takes about twelve days. Cells from which wasps have hatched out are often used over again. Combs may therefore exhibit zones containing brood of different ages, moving out from the centre as time passes.

At the height of the summer, males and females begin to be produced instead of workers. What starts this off, is rather obscure. Obviously the later the wasps leave it the more workers can be produced and these will be able to rear more queens and males. On the other hand if it is left too late, there may be no time to produce a good 'crop' of breeders before the frosts come and cut the colony down. Queen cells are recognizably larger with some species, for instance *V. vulgaris* and *V. crabro*; with other species they do not appreciably differ from worker cells. There is no evidence that larvae of queens-to-be are given food different in quality from that for worker-larvae. It may, however, differ quantitatively. As with many other hymenoptera, unfertilized eggs give rise to males; how the queen arranges that males and females are produced in the right proportions is a matter to be discussed more fully in chapter 24 on honey-bees.

Mating of males and young queens takes place outside the

nest. In the case of flying, that is, male and female ants, there are often assemblies on mountain-tops or other conspicuous high places where mating takes place. This is not usually seen among bees or wasps. On one occasion, however, I saw very large numbers of mating wasps on the top of the Prescelly mountain in South Wales. All round the top of the mountain were little balls of struggling wasps, a queen in the middle and a lot of males all trying to mate with her. Unfortunately I did not realize at the time that this was an unusual occurrence and so did not identify the species. A similar incident was reported, also in South Wales; in this instance *D. sylvestris* was the wasp concerned.

Towards the end of the season, a wasp colony goes into decline. *Dolichovespula* colonies may even do so in mid-summer though insect food is still plentiful. In declining colonies there may be cannibalism. Some larvae may be killed and used as food for others, or thrown out of the nest altogether. Possibly it happens when there are too few workers to cope with feeding a lot of queen and male larvae. Quite apart from this both queens and workers engage at times in egg-eating; this, too, may be a sort of birth control practised when there is an imbalance between numbers of workers and of larvae requiring food. As the weather gets colder, fewer workers are seen and these are increasingly sluggish; they, and not only queens intent on hibernation, are often found indoors.

Wasps commonly do not sting people except in self defence or when they or their nests are interfered with. There is some evidence that wasps in very populous colonies are particularly fierce. This may be related to the greater aggressiveness which is associated with higher temperatures. The interior of a wasp colony is often 5°—15°C. hotter than the surrounding air: in very populous colonies the larger numbers of wasps would tend to raise the temperature rather more. A wasp's sting has has only microscopical barbs; so it is not held in the skin as a honey bee's sting is: it can be used for repeated thrusts.

Otherwise it is just about as unpleasant as a bee-sting, but is only dangerous for the very small number of people unfortunate enough to be highly allergic. As already mentioned, bad weather in spring forms a serious hazard for queens. Another limiting factor is a scarcity of suitable nesting-sites; accordingly many fights ensue between queens which have found a site and usurper queens. Such fights often lead to the death of one queen or even both. There are also *Vespula* species which are cuckoos.[14] *V. Austriaca* is a species parasitic on *V. rufa* in both Europe and America: in Britain it is commoner in Scotland than farther south. The American *Dolichovespula adulterina* similarly parasitizes *D. arenaria*. *V. squamosa* sometimes makes its own nest, and is sometimes a cuckoo in *V. rufa* nests. The cuckoos have to find a wasp colony already going well and with enough workers to rear the young they propose to introduce. Apparently they are accepted into the nest without a struggle.

Man, also, may be an enemy to wasps. He cannot be blamed for destroying the wasps which come into the kitchen or buzz round the food, especially if he is eating out of doors. He should not, however, forget the numbers of other insects, many of them harmful, which can be destroyed by a nest containing perhaps 5,000 wasps. In any case he cannot expect to influence matters greatly, except perhaps by destroying active nests. In Cyprus, a local hornet species was being troublesome. A reward of rather less than a farthing was offered for each queen destroyed and in due course over 200,000 were paid for: next year, nevertheless, there were far more of the hornets about than usual.

Though they have many enemies, some insects are their friends. Larvae of hover-flies of the genus *Volucella* are commonly found in wasp's nests, but they are scavengers, as are other species in bumble-bee nests (see p. 159). The wasps do not seem to hinder the going in and out of the flies. Fabre[7] records that their larvae push past wasp larvae in cells and seem to clean up behind them. He suggests that the Volucella larva performs two useful offices: 'she wipes the wasp's children and she rids the nest of its dead'. Two *Volucella* species have strikingly banded abdomens, somewhat resembling those of wasps. It seems improbable that this resemblance confuses the wasps, so that they allow the flies to enter: it is far more likely that this is a case of protective resemblance, discouraging birds and other predators.

Species mentioned

Vespa crabro	BA	Vess'-puh cray'-broh
Vespula vulgaris	B	Vess'-pu-luh vul-gayr'-is
— germanica	BA	— jermann'-i-cuh
— rufa	BA	— roo'-fuh
— pennsylvanica	A	— penn-sil-vann'-ic uh
— austriaca	BA	— auss-try'-ac-uh
— squamosa	A	squay-moh'-suh
Dolichovespula sylvestris	B	Doll-i-ko-vess'-pu-luh sill-vest'-riss
— adulterina	A	Doll-i-ko-vess'-pu-luh add-ulter-ine'-uh

Scavenger

Volucella	BA	Voll-u-sell'-uh

CHAPTER 15

Comparative Studies of Behaviour

THE preceding chapters have described the different behaviour of many kinds of wasp, as regards the sites of their nests, the nature of their prey, the way they carry it, the way they provision their nests and lay their eggs. Many of the facts are in themselves quite remarkable or even entertaining. What is, however, of special interest is to compare the habits of all these species and try to learn what light such study can throw on evolution; this is the province of comparative ethology.

At this stage we shall consider only the differences between the various wasps, to see how evolution among them has progressed. Discussion of broader problems – the origin of aculeates, the emergence of the social structure and the development of parasitic habits – will be left to the final chapter of the book.

Those who attempt to classify wasps, and other insects too, rely almost entirely on structural characters, the form of their bodies and legs, the arrangement of their wing-veins. These are readily observed and, indeed, are the only things which can be observed in the museum specimens which form the material upon which the taxonomist works. It can happen, however, that two populations of wasps, alike in appearance, have striking differences in behaviour; these, it can be argued, should be just as useful criteria for dividing species as small differences in structure.[38] A classical example is to be found in the genus *Ammophila* (chapter 4).[25] In some places it was found that an *Ammophila* (*A. campestris*) was provisioning its nests with caterpillars of moths. Elsewhere, but not in precisely the same areas, it was hunting only sawfly larvae. Following this clue, Adriaanse found that there were two species concerned, and when they were studied with sufficient care, it became possible to detect also some previously overlooked differences in structure. The caterpillar-hunting one has since been given the name *Ammophila pubescens*. In a similar way Drs. Evans[38] and Krombein were separately studying two pompilid wasps *Anoplius semirufus* and *apiculatus* so similar as to be distinguishable only with a strong lens. Both species nest in bare sand and carry their spider backwards, holding it at the base of the hind legs. They dig their burrows similarly, hunt and store wolf-spiders (*Lycosidae*) in a similar way and the duration of the developmental stages is the same. Though both nested in bare sand, one did its hunting there, but the other went to adjacent woods and therefore caught different Lycosid species; one of the two constantly vibrated its wings when hunting, one visited flowers for nectar, the other sucked juices from its spiders: and so on, with other minor differences.

Comparative ethology may also give us clues suggesting that current ideas about classification may not be correct. This has been emphasized by H. E. Evans,[35] who has reinforced his ideas with facts concerning the structure of the larval forms of wasps. He suggests, for example, a fresh look at the *Sphecinae*, the sub-family of the big Sphecid family which includes *Sphex* and the related wasps considered in chapters 3, 4, and 5. These differ in larval characters from the other sphecids, and, alongside that difference, they pack down the nest closure by blows of the head, while the other sphecids use a hammer-like structure on the tail. Further, the fly-catching genus *Mellinus* (chapter 10) differs considerably from the wasps with which it is commonly classed (*Nysson*, *Bembix* and others), not only in the structure of its larvae, but in a number of its habits.

In trying to observe the course of evolution, we find that morphology and ethology do not always go hand in hand. Some genera which have not diverged very far from their ancestors in their structure, have yet evolved in the direction of very complicated, specialized habits; and the converse is equally true.

There has doubtless always been intensive competition, so that wasps, like other animals, have had to specialize in order to take advantage of any 'ecological niche' not already occupied. Evans' CCCC principle, that complete competitors cannot co-exist, has already been mentioned. We can thus see how different genera have come to exploit different sites for nesting and different kinds of prey. Other differences in habits must often have come about as secondary consequences of these major divergences.

In chapter 2 it was pointed out how disadvantageous it was for the members of the relatively primitive pompilid family to have to drag their prey backwards along the ground, so that they could not see where they were going; and to hunt before digging a nest-hole, so that they had to leave their prey in the open at the mercy of ants or other thieves or parasites. *Scolia*, one of the most primitive of wasps, stings its prey and oviposits *in situ*, as do some pompilids and other primitive wasps: this is a relic of their probable ancestry among the parasitic hymenoptera.[61] These wasps leave a single paralysed victim for their offspring. With this legacy, the pompilids have been rather tied down to their existing ways of transporting and provisioning. With the development of the habit of storing numbers of smaller victims in each cell, transport at once became an easier matter. The kind of prey used also gives some clue to evolutionary history. The more primitive scoliids, pompilids and *Sphecini* prey on types of arthropods which also have a very long history—beetles, spiders and grasshoppers. Many of the more specialized sphecids hunt flies and other hymenoptera, themselves more recent products of evolution.

Carrying the prey in the jaws seems to have been the original method of transport; later legs came to be used, and on the whole a later tendency has been to use middle or hind legs rather than front ones, the transport mechanism moving always backwards. The centre of gravity of the prey thus comes to be closer to that of the wasp, making the transport easier and quicker. The final stage is reached when we find *Oxybelus* carrying home its flies impaled on the end of its sting. There is even a further refinement: *Clypeadon*, which preys on ants, has a specially modified tip to its abdomen which neatly fits in between the bases of the ant's front and middle legs, forming an efficient 'ant-clamp'. Evans[32] has published a diagram, showing how the change from jaws to legs to tail as transport mechanisms, may be correlated with other characters which indicate the extent of evolution and specialization.

One advantage of carrying the prey more and more posteriorly is that it allows wasps which practise it to keep their front legs free. Thus they can close their nests when they leave, and the front legs will be available when they return, to clear away the temporary closure so that the wasp can enter without letting go of its prey. There is thus less chance for the omnipresent parasites. Other devices for thwarting parasites are practised by the pompilids which temporarily bury their prey or 'park' it above ground where ants are less likely to find it. False burrows are made by a number of different sphecids,[37] while others go to great trouble to level off the ground near the entrance to their hole or otherwise camouflage it. It seems a little odd that when so many devices such as these have been developed, there is so little tendency for wasps to attack and kill their parasitic enemies; the miltogrammine flies in particular, would seem quite defenceless and they are more universally troublesome than any.

The habit of progressive provisioning seems to be an advanced development with several advantages. It is suggested that it arose when wasps had partly finished provisioning a cell and were interrupted by bad weather or onset of darkness, so that the task had to be finished next day. Some wasps, indeed, do this, even though they are not regular progressive provisioners. The classical progressive provisioners, the *Bembix* species, can stay on guard when not actually hunting and can keep away parasites; yet their usual method of devoting all their time to one larva over several days must make it a slow job to care for numerous progeny one after the other. The plan adopted by *Ammophila pubescens* of looking after several nests at once holds much more promise and must have been what happened on the road of evolution of the social wasps.

Species mentioned

Anoplius semirufus	A	Ann-oh'-plee-us semmy-roof'-us
— apiculatus	A	— ay-pick-u-late'-us
Clypeadon	A	Kly-pee'-ad-on

Prey

| Lycosidae | Lie-cosy'-dee |

CHAPTER 16

Bees, and Especially Short-tongued Bees

THE bees form a large and successful group of insects, probably arising long ago from ancestors to be found among the sphecid wasps. The key to their success doubtless lies in their use of honey and pollen for feeding their larvae; these substances, being universally available in flowers, must form a much more dependable source of provender than insects or spiders which may be hard to find, catch and transport.

A very interesting study in the field of evolution is to be found in the parallel adaptation of flowers to insects and insects to flowers. Bees are particularly important in cross-pollinating flowers since they need to spend some time, perhaps deep in a flower, collecting pollen as well as nectar. Some are described as 'wallowing' in a flower. Moreover their apparatus for transplanting pollen is just what is most useful to the plant which needs to have its fertilizing agent moved around. Most other insect-visitors are merely in search of a little nectar for their own nourishment and their relatively restricted activity may be comparatively useless for cross-fertilization. Many plants have, accordingly, in the course of evolution, developed modifications of structure making them particularly attractive to bees, or so hiding their nectar that only relatively deep penetration by a bee and its proboscis will avail to procure it. There are many instances in which the mutual adaptation of bee and flower has gone so far that a species of bee may practically confine its attention to a particular kind of flower. In Britain *Melitta tricinta* collects practically all its honey from a wayside flower, red Bartsia (*Odontites verna*) while another member of the genus, *Pseudocilissa dimidiata*, only recently discovered in one British locality, is found only on a papilionaceous flower, sainfoin (*Onobrychis sativa*), often

8

cultivated for fodder. There is often an association between the length of a bee's proboscis and the depth of the tubular corolla of the flower which it particularly frequents. Knowledge of what flowers are visited by bees is sometimes of importance to agriculturalists or horticulturalists, concerned about obtaining a good yield of seed from their crops. This can be ascertained by microscopical examination of the pollen the bees collect.

Bees differ from wasps in a number of structural characters. Most noteworthy are two which represent adaptation for collecting pollen. There are, at least in the females, dense hairy areas, usually on the legs, sometimes under the abdomen, specially intended for pollen collection. In nearly all bees the basitarsus of the hind legs (Fig. 14, p. 153) is widened and often covered with short hairs, and the tibia is either very hairy or, in social bees, smooth and bare but surrounded by curved hairs to form a pollen-basket. Further, the hairs of bees, unlike those of wasps, are plumose, that is feathery, or branched. In some parasitic species which no longer collect honey, the branched hairs are very few and it may only be possible to find a small number at the front of the thorax. In the bees with pollen-baskets the pollen is moistened with nectar to make a sticky paste: those with hairy legs collect the pollen in a dry state. Most bees, when out foraging, take a certain amount of pollen for personal nourishment.

The nesting habits of solitary bees are on the whole very similar to those of wasps. Some make their nests in dead wood, or tunnel in bramble or other stems; even more species tunnel in the ground. They seem on the whole to have developed fewer devices for thwarting parasites than have the solitary wasps. Many leave mounds of earth outside their burrows, not attempting to camouflage the entrance as some wasps do, by scraping away sand, nor making false burrows or even temporary closures.

Among the most primitive bees are those of the genus *Hylaeus* which are small and black usually with some yellow or white markings on the face or legs. Their tongues are quite short and concave at the tip and, unlike other bees, they have no elaborate devices for carrying pollen. At most there are some longish hairs on the front legs of the females, useful when pollen is being collected. They imbibe nectar, only being able, of course, to obtain this from flowers with a relatively shallow corolla. Favourites for visiting are bramble, mignonette and umbellifers such as wild carrot. The nectar and pollen are carried back to the nest in the bee's crop or honey-stomach and there regurgitated.

Cells are placed usually in a single series in any suitable cavity. Most frequently perhaps they are in hollowed-out stems of brambles or other plants. Partitions are made of chewed up pith, though with some species there are hardly any partitions at all between cells or very flimsy ones.[45] The cells are lined with delicate membranes made from the bee's salivary secretions. The shape of the bee's short tongue may bear a relation to the application of this material to the nest-wall.

Cells may also be built in abandoned nests of mason-bees (chapter 20) or in old earthworm burrows. The stored food is semi-fluid, the egg is laid on the surface, but the larva may subsequently be found almost buried in the glutinous mass. They are thus simple bees, leading apparently uncomplicated lives. Not many parasites are recorded for them: in Hawaii, however, there are some *Hylaeus* species which have become parasitic on other members of the genus. *Hylaeus* bees are said to emit a faint perfume.

The other important genus of short-tongued bees is *Colletes*. These are much larger and hairier bees than *Hylaeus*, rather like small honey-bees but with bands of closely-appressed short white hairs on the margins of the abdominal segments of most species. There are pollen-brushes on the hind legs. One British species, *Colletes succincta*, particularly frequents sandy commons and visits heather flowers; the other species are largely coastal and get their honey from composites such as ragwort (*Senecio*) and chamomile (*Matricaria*). They tend to nest in colonies which may contain hundreds of burrows close together; these are often in rather hard sandstone but may be in loose sand. As with *Hylaeus*, the tunnels are lined with a water-proof membrane formed from a salivary secretion; it has been likened to gold-beater's skin. The largest species, *C. cunicularis*, is in Britain only found on sand-hills in Lancashire and Cheshire. Unlike the other species it is a spring bee. It digs burrows 12 cm deep when in hard soil but as much as 28 cm long in loose sand.[53] The food stored is, again as in *Prosopis*, more liquid than with longer-tongued bees. Very often it goes sour or mouldy, but this seems not to matter; the larvae still thrive on it. The bee's egg is glued to the roof or the ceiling of the cell instead of being placed directly on the food store. When it hatches — not for several weeks in the case of *C. cunicularis* — the larva is, according to Maly-shev[53] 'almost imperceptibly (apparently passively)' trans-ferred to the provisions. Thereafter it hardly moves but sits and eats honey, forming a progressively enlarging hollow round itself. As with many wasps the excreta are not voided till the feeding is finished.

The special parasites of this genus are species of *Epeolus*.[58] Most cuckoo-bees are members of genera closely related to their hosts and doubtless derived from them in the not too distant past. Not so *Epeolus*; they are rather dumpy short-haired, long-tongued bees, strikingly marked with black and white and with some red on the legs (Plate 10B). It is an interesting matter to speculate how such a very different bee came to specialize in parasitizing the quite unrelated longer-haired brown *Colletes*. Possibly the ancestor of *Epeolus* was a working bee with habits similar to those of *Colletes*, taking to parasitizing its neighbours in a way to be discussed later (p. 186). And then perhaps the honey-gathering ancestor died out, leaving its parasitic offshoot surviving as a hanger-on of *Colletes*. Be that as it may, an *Epeolus* has been found in up to a third of the cells of one *Colletes* species. The larva kills the egg or very young larva of its host: later it has a remarkable flattened form and eats the whole food store.

Species mentioned

Melitta tricincta	B	Me-litt'uh try-sink'-tuh
Pseudocilissa dimidiata	B	Pseudo-si-liss'uh di-middy-ate'-uh
Hylaeus	BA	Hi-lee'-us
Colletes succincta	B	Coll-ee'-tees sux-sink'-tuh
— cunicularis	B	— cu-nick-u-lair'-is
Epeolus	B	E-pee'-oh-lus

CHAPTER 17

Some Mining Bees

MANY bees besides *Colletes* make their nests in underground burrows. Among the most numerous are the species of *Andrena*, of which there are over sixty in Britain and far larger numbers elsewhere. (Plate 11A.) The largest are as big as honey-bees, but most are rather smaller and some only 6 to 7 mm long. Most of them are very hairy, some are conspicuous insects and a considerable number have pale bands

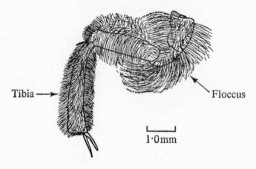

Tibia →   ← Floccus

1·0mm

11. Leg of *Andrena*

on the abdomen, made up of fringes of hair. This banding seems not to give a sufficiently wasp-like appearance to be likely to deter predators, but it is so prevalent that perhaps it has such an effect. Pollen is carried on the densely-haired hind tibiae of the female, but in some species there is in addition, at the base of the hind legs, a collection of curved hairs, the floccus, having a similar function (see Fig. 11). As is often the case with genera containing many species, there is often difficulty in distinguishing between members.

Many *Andrena* species are among the earliest spring bees and may be found collecting nectar and pollen from sallow bushes, dandelions and other early flowers. The males may be seen coursing up and down hedges, often several species together, searching for females. Dr. R. C. L. Perkins,[55] writes that these bees, even those of local species, may be so numerous that several can be caught 'with a single stroke of the net'—yet all will prove to be males. Many species burrow in sand, and around colonies these may be exclusively females. These colonies may extend for several hundred yards. The bees are are of course 'solitary' in the sense that each female works only for her own offspring; yet there are the first signs of evolution towards colonial life in that in a few species, as with a few wasps, several females will share a common entrance burrow, making their own homes in side burrows arising from this. The rather local species with large-headed males, *A. bucephala* and *A. ferox*, do this. There are, on the other hand, numerous Andrenas nesting well apart from any others. A large colony of *A. humilis* was found in Gloucestershire in 1876 and was still there nearly forty years later. Perkins writes 'As no other colony of this bee has ever been observed in the district, we may assume that this one has maintained its hold for forty years at least, and might conceivably have existed for centuries'.

The earth or sand from excavation of the tunnel forms a little mound at the entrance. *A. armata* is a common British bee of which the female is densely covered above with thick orange-red 'fur'. Mounds round its burrows are commonly seen on garden lawns and are regarded by those not interested in insects as untidy. Another species (*A. vaga*) builds at the burrow's entrance a sort of turret which is concealed by the mound round it.[53] Cells may be in a linear series or may be set at angles to the main shaft; they may be lined with a silky material. The food stored in the cells is in the form of a loaf of pollen stuck together with nectar.

Both sexes come out of their nests in spring or summer. At least some spring species hatch out from the cocoon even earlier and wait patiently underground till it is time to come forth. On two occasions I have, when gardening in mid-winter, dug up fresh females of the spring bee *A. haemorrhoa*. Some Andrenas are double-brooded, while some are single-brooded further north and double-brooded in the south. There may be structural or colour differences between those of first and second broods. It is quite strange that of two spring species which are active at the same time, one alone will go on to produce a second brood. A few species restrict their honey-gathering to one kind of flower, or to one or two closely related ones: *A. florea* to white bryony (*Bryonia*), *A. cingulata* to speedwell (*Veronica*). Several species seek to evade capture by feigning death: they lie quite still with legs closely applied to the body and may thus passively reach the bottom of vegetation where they may be hard to find.

Almost regularly associated with species of *Andrena* are members of the parasitic genus *Nomada*, wandering bees (Plate 11B). These are very different from *Andrena*, all being relatively hairless, strikingly banded or otherwise marked with black and yellow or red. (Fig. 12.) They look, in fact, very wasp-like and this appearance is doubtless effective in deterring predators. Such deterrence is very necessary, for

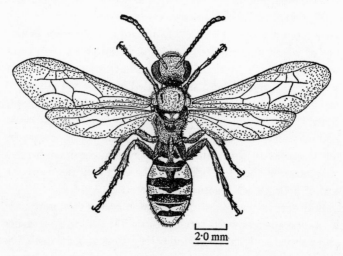

2·0 mm

12. *Nomada marshamella*

the females spend most of their lives in the open, seeking an opportunity to enter and lay in *Andrena* nests. Of the many *Nomada* species, most parasitize a single *Andrena* species or one or two closely related ones.[55] Their conspicuous habits are such that in a locality one may well chance to see many more Nomadas than Andrenas; the former are out busily looking for nests of their hosts; the latter are often occupied underground. One may even discover that an *Andrena* is present locally, only be catching the corresponding *Nomada*!

When they first hatch, Nomadas do their resting in the burrows within which they were reared, but as soon as their hosts become active in nest-building, they sleep out, clinging to the heads of grasses and there looking rather like an extra seed-head. Like some parasitic wasps and other parasitic bees they have a hard thick cuticle, but it is uncertain how far there occur actual fights between the cuckoos and their hosts. At any rate Nomadas mostly play safe, by waiting till a female *Andrena* is away before entering her burrow. Some Andrenas when not out collecting food will guard the entrances to their burrows to keep the cuckoos away.

Nomada is placed in a different family of bees from *Andrena*, though it is not as distant from it as *Epeolus* is from *Colletes*. One can with fair reason maintain the view that at some remote period the two genera had common ancestors and that the stock divided into those which worked for themselves and those which were parasitic. Thus as the *Andrena* genus evolved to give rise to its many species, Nomadas evolved in parallel fashion, each member adapting itself to its own host. This idea is supported by the fact that, on the whole, related species of *Nomada* are associated with Andrenas which are also closely related to one another.

The following chapter will deal with another genus of mining bees, *Halictus*; the special parasite of these is *Sphecodes*, a closely related genus. There has been a little transfer of parasites between these two groups of genera. Thus, one of the smallest Nomadas (*N. sheppardana*) is associated with *Halictus nitidiusculus* and probably other small *Halicti*. On the other hand, one or two *Sphecodes* have transferred their parasitism to species of *Andrena*. I remember, more than fifty years ago, watching *Andrena barbilabris* on Hampstead Heath in north London diving down and finding its nest in loose sand— most mysteriously since this must have had its conformation altered all the time by passing walkers. Plunging down at the same spot was *Sphecodes pellucidus*; I thought I was the first discoverer of the association of a *Sphecodes* with an *Andrena*: this proved, of course, to be known already though it was not in my book. Such an association reveals an interesting adaptation. *Halictus* and *Sphecodes* are among the genera in which emergence and pairing take place in the autumn, and only the fertilized females survive the winter, ready to start nest-building in spring. On the other hand Andrenas of both sexes come out together and pair in the spring or early summer. *S. pellucidus* and another (*S. rubicundus*) which parasitizes an Andrena (*A. labialis*) have abandoned the practice of their relatives and have adopted the same life-history time-table as their *Andrena* hosts.

Andrenas of several species are parasitized, as wasps may be, by species of bee-fly, *Bombylius*. The common British *B. major* is often seen hovering over primroses and other spring flowers. A distinguishing character is that its long non-retractable proboscis sticks out permanently in front of it.

Very common parasites are insects of the genus *Stylops*. These and a few related genera have been placed in an order all their own, *Strepsiptera*, but they are now considered by many to be strangely modified beetles. The males, only a few millimetres long, are rarely seen. They fly around with the aid of triangular lower wings; their fore-wings are reduced to mere rudiments. This of course is in harmony with what is seen among beetles, where the hind-wings do all the flying. The larvae are internal parasites. When they have finished feeding, the pupa is extruded between the dorsal segments of the bee's abdomen, most often between the fourth and fifth.

When the time is ripe the pupa-case bursts and the male *Stylops* shoots out and flies around seeking the female, but only for two or three hours: he then dies. The female, on the other hand, never leaves the body of the bee; just her head and thorax project between the bee's abdominal segments. A number of male *Stylops* may be seen assembling round a female. Perkins[56] describes an assembly of 'not less than fifty . . . a branch appeared quite white from the wings of the small creatures'. Within the body of the fertilized female are formed nine to ten thousand minute larval forms, known as triungulids. They emerge from three to four weeks after fertilization and somehow reach the bee's nest. It is known that an *Andrena* can take triungulids into its alimentary canal along with nectar and regurgitate them on to the pollen-mass in the cell. The triungulids are able to penetrate the bee's egg and have later been found floating inside the larva. They do not apparently seriously begin feeding till the larva has reached the prepupal phase of development.

This internal parasitism is not without its effects on the adult bee. The reproductive glands are particularly affected and 'stylopized' female bees seem never to lay fertile eggs. The male bees are on the whole less affected and may mate with normal or stylopized females. Male *Stylops* appear to affect the bee's development more than do the female parasites. The damage to ovaries and testes has the result that the bees lose their secondary sexual characters and so the sexes come to resemble each other. Females become less hairy and in particular the pollen-collecting hairs on the hind tibiae are poorly developed. In some Andrenas the male has a white and the female a dark face. As a result of stylopization the female may develop white on the face while the male's face may become dark. One effect of all this is that changes occurring among closely related species may make it hard to determine which is which: moreover bees changed by this parasitism have been described as new species.

A related genus of parasites has similar effects on species of

Halictus and solitary wasps may also be afflicted, though not in Britain.

There are a number of other genera of mining bees, closely related to *Andrena*. Species of *Melitta* attached to particular flowers have already been mentioned. *Eucera* is a striking genus, the males having extraordinary long antennae. *Dasypoda hirtipes* is a very handsome coastal species of which the females have hind legs clothed with remarkably long shaggy hair. Their habits, however, require no special mention. We therefore turn to *Anthophora*, mining bees of a different family from *Andrena*: at least one species nests in wood rather than in the ground.

Two species of *Anthrophera* (*A. retusa* and *pilipes*) are common in Britain in the spring. They have rounded furry bodies like a bumble-bee's but rather smaller. The females are jet black and the males rusty brown. Some of them have curious arrangements of hairs on the middle legs (Plate 12A). They have a characteristic flight, hovering motionless in front of a flower, then dashing rapidly off to do the same elsewhere. Entrances to burrows may have a protective archway and the cells are lined with a sort of glaze. These bees are particularly industrious, rivalling the honey-bee in the long hours of work. This may be necessary because they provide their brood with very liquid provisions, which must need very many journeys for transportation.

They are parasitized by a closely related genus of bees, *Melecta*, black bees with a few white markings. According to one account,[8] *Anthophora* politely stands aside to let the parasite into its nest; others, however, have seen the two species engaged in a struggle.

Their enemies also include the beetles *Melöe* and *Sitaris*, which have a strange life history. *Melöe*, a large clumsy black beetle, lays its eggs rather at random in very large numbers, perhaps 10,000. *Sitaris*, studied in France by Fabre, has a similar life history. The hatched larvae, called triungulids like those of *Stylops*, hibernate in a heap and in the following

spring get on to flowers and thence attach themselves to any
hairy insect. A few of the thousands of them will be fortunate
enough to attach themselves to an *Anthophora*. This will most
often be a male, for these come out well ahead of the females;
but the triungulids manage later to transfer themselves to
the latter. In due course the female *Anthophora* fills her cell
with honey and lays her egg on top. The triungulid glides from
the body of the bee on to the egg and remains perched on it as
on a raft, floating on the honey. It spends about eight days
consuming the egg and then moults, turning itself into the
form of a little bladder with the spiracles for respiration on
top. In this form it is well fitted to float on the honey, which is
consumed in about forty days. It next turns into what is called
a pseudo-pupa. Some pseudo-pupae go through further changes
resulting in the emergence of adult beetles in August or
September. Most, however, winter as pseudo-pupae, and in
the following June turn once more into a larval form not unlike that of an earlier stage. This does not shed any coverings
but remains within the two dead skins for a few days, then
turns into a genuine pupa from which it duly hatches about
a year later than those which had come out without all this
delay.

Species mentioned

Andrena bucephala	B	An-dree-'nuh bu-seff'-a-luh
— ferox	B	— fer'-ox
— humilis	B	— hu'-mill-is
— armata	B	— ar-mate'-uh
— vaga	B	— vay'-guh
— haemorrhoa	B	— hem-o-roh'-uh
— florea	B	— flor'-ee-uh
— cingulata	B	— sin-gu-late'-uh
— labialis	B	— lay-bee-ail'-is
— barbilabris	B	— bar-bee-lay'-bris

Species mentioned (*cont.*)

Nomada fucata	BA	Noh-'mad-uh few-kate'-uh
— sheppardana	B	— shepp-ard-ayn'-uh
Eucera	BA	You'-ser-uh
Dasypoda hirtipes	B	Day-si-poh'-duh hirt'-i-pees
Anthophora retusa	B	An-thoff'-or-uh re-tyus'-uh
— pilipes	B	— pie'-lip-ees
— acervorum	B	— asser-vor'-um
Melecta	B	Me-lect'-uh
Halictus nitidiusculus	B	Ha-lick'-tus nite-iddy-us'-cu-lus
Sphecodes pellucidus	B	Sfeck-oh'-dees pell-loo'-sid-us
— rubicundus	B	— ruby-cund'-us

Parasites

Bombylius major	Bom-billy'-us major
Stylops	Sty'-lops
Meloe	Mell-oh'-ee
Sitaris	Sit'-a-riss

CHAPTER 18

Halictus: Beginnings of Social Life

'DO YOU know the *Halicti*? Perhaps not.'—so writes Fabre[9]. 'There is no great harm done. It is quite possible to enjoy the few sweets of existence without knowing the *Halicti*.'

Bees of the genus *Halictus* are many of them small and black, usually smaller than *Andrena*; some of the larger species, however, have pale abdominal bands and look much like Andrenas. Some European species have a slight metallic tint but a number of American ones are bright metallic green. I recall that on my first visit to America, over forty years ago, I was much excited by the strikingly coloured *Halicti* on the asters and golden-rods. The females may be recognized by the presence of a bare strip at the end of the abdomen, on the dorsal surface. As was mentioned in chapter 8, on bee-wolves, *Halictus* has now been subdivided into a number of genera, of which most are American. It is convenient, however, to use the more familiar name for those now considered.

The genus is of particular interest, since one can trace within it gradations from a completely solitary habit to a definitely social life. This aspect has been of interest to several European observers including Fabre; and in America 'the nest-architecture of the sweat-bees (*Halictinae*)' has been exhaustively studied by Sakagami and Michener.[21] Though the term 'sweat-bees' is used by these authors, it will not be used here, as it is more appropriately applied to the *Meliponinae*, the stingless bees of the tropics; these really are troublesome when they dance around one in clouds, as flies of the genus *Hydrotaea* do further north.

Nearly all *Halicti* are miners in the ground, only a few nesting in rotten wood. They particularly like steep banks and many of the little bees may be seen flying in a zig-zag manner up and down on such a bank, which may be riddled with holes. Fabre estimated that there were a thousand nests of *H. calceatus* in an area of 10 square metres, while a figure of 3,500 nests is quoted for a mixed colony of two other species. Aggregations of twenty or thirty nests are, however, more usual. One colony of *H. malachurus* has been found at the same spot over a period of thirty-five years. Several species may be found nesting together in a bank. Nests of a predatory *Cerceris* have been found right in the middle of a *Halictus* colony; this is perhaps not as bad for the *Halictus* as it seems, as this *Cerceris* did nearly all its bee-hunting on flowers away from the colony.

Some of the species have life-histories like those of *Andrena*; a difference is that both sexes emerge in late summer and pair then; only the fertilized female survives the winter, starting to rear her family in the spring. Some species have two sexual generations a year. Much more interesting are those which have modified their habits towards a social life. The species most studied has been the European *H. malachurus*. In the south of France the hibernated females start digging their burrows in February: in Britain this does not happen until April; many miniature mole-hills can then be found within a small area. Sometimes there is co-operation between several over-wintered females; these make their own cells, though using a common main burrow. Fabre[9] describes the cells of *H. zebrus* as like vaccine-phials laid on their sides and opening into the passage. They are lined with a glaze of water-proof material made from saliva; if the earth is washed away, the delicate lining is left. Not all *Halicti*, however, make such a lining. The cells are duly stored with honey which is made into a little cake as with Andrena, and the egg laid on top. Progressive provisioning occurs in a few social species. The bees which hatch out from the early *malachurus* nests are all females. These, though looking like their mother are, in fact, workers. The situation is much as among the paper-wasps described in chapter 13. These workers now proceed to build and store

9(A) *Dolichovespula sylvestris* at the entrance to its nest

9(B) Wasp-nest broken open to show cells and canopies

10(A) *Polistes nimpha* on its comb

10(B) *Epeolus cruciger* investigating a *Colletes* burrow

more cells, enlarging the nest in which they were born. The queen, the foundress of the colony, may continue to forage but she is more often found acting as a portress her head closing the entrance to the burrow. (Plate 13A.) When one of the workers returns with provisions she withdraws to let her enter, but she is soon back again to prevent the entry of enemies. When the colony has reached a certain size, towards the end of summer, the queen begins to lay eggs which will produce males and potentially fertile females. These will mate and the females will hibernate to continue the cycle. In some species pairing commonly takes place underground, for mating couples have been dug up. Others pair on flowers outside. In either case the males are seen in plenty in the late summer or autumn, taking their fill of nectar from flowers of many kinds (Plate 13B). In the case of *H. malachurus* there are small differences between the first and second generation females such that these were at one time thought to belong to different species.

The history just described is more characteristic of old world than of American *Halicti*. Fabre,[9] who was in general such an accurate observer, watched his colonies of *malachurus* but unfortunately misinterpreted what he saw. He realized that the hibernated queen produced at first only females but supposed that these then gave rise by parthenogenesis, to the later generation of males and females. As a result of study of the habits of various European species it has been suggested that the evolution of the social habit has proceeded along the following course. First, there are solitary bees, having one generation a year. Then there are two generations produced. Next the hibernated female has two breeding periods in the summer, the first generation rearing their own broods: autumn bees would thus some of them be children and some of them grandchildren of the original female. A later development would be that all the female offspring remained with their mother, perhaps producing males parthenogenetically, but later ceasing to do so. It would then be a short step to the

9

position seen in *malachurus* and its relatives in which the queen produces at first only females destined to be infertile workers. In one species (*H. marginatus*) in Europe a colony has been thought to survive with the orginal queen up to five years and to contain as many as 700 cells with 300 adult bees looking after them at any one time.

Further resemblances to the behaviour of the social bumble and honey-bees are apparent when we consider the evolution of the nest architecture.[21] The tendency to build a comb has apparently developed, however, independently of any changes in social organization. There is always, and in contradistinction to the habit of most fossors, a mound of earth around the burrow entrance, at least until it weathers away. There may also be an entrance turret—up to 47 mm high in a South American species; it may gradually increase in height in colonies surviving for more than one year. The mouth of the burrow is narrowed, presumably to make the task of the portress easier. One American species (*H. humeralis*) makes an oblique burrow and a concealed second entrance to the main burrow, much as was described for *Bembix pruinosa*. The main burrow may go down for 70 mm (*H. malachurus*); there is nearly always a blind burrow at the bottom of the main shaft.

One Australian species makes burrows leading to a single cell, but nearly all others make branched nests; only a few have their cells in a linear series. The cells may open directly into the main burrow or be found at the ends of side passages; these in turn may lead to groups of cells, variously arranged. Finally there are numerous species in which clusters of cells are grouped to make a little comb. *H. malachurus* is one bee which constructs combs at times, but the best comb-builder is *H. quadricinctus*, a European species very rare in Britain. In combs the cells are most commonly horizontally placed, walls between individual cells being quite thin. The clump of cells is usually surrounded by a cavity; this is thought to assist ventilation and to make it easier for the queen to keep the cells warm by brooding or cooler by fanning. A cover may be built over the comb. The individual cells are first made rather larger than is required, then narrowed and the walls made smoother by lining with clay and perhaps salivary secretions. The growing larvae do not retain waste material till they have finished feeding as do most other wasps and bees; but what is voided earlier is stuck on to the walls of the cell to keep it out of the way of the food.

Needless to say the *Halicti* have their parasitic enemies. Fabre has described the tactics of a white-faced miltogrammine fly; this watches and waits while the *Halictus* is working. The egg must be laid at just the right moment when the storing of the food is finished, for the bee's final act before laying is to knead the pollen-nectar mixture into a ball. A prematurely laid fly egg might well be destroyed during this process. Two or three fly larvae eat all the honey-store and the bee larva starves to death. The parasites pupate outside the cells and do not hatch till April, though the fly had laid in nests in the summer. Thus there may be only one generation of the flies during two generations of bees.

Sphecodes is the special cuckoo-bee afflicting *Halictus*, as in the last chapter. The name means Sphex-like or wasp-like and is not particularly appropriate. Most *Sphecodes* are black with the abdomen largely red. A red colour in *Halictus* is usually seen only in the male sex. A criterion of importance in differentiating *Sphecodes* species is in the number of tiny hooks joining the front and hind wings together in flight. Most species of *Sphecodes* attack one or at most a few closely related *Halictus* species. Their method of operation differs little from that of other parasitic bees and wasps. They are said to destroy the *Halictus* egg before they lay their own. The portress *Halictus* queen guarding the nest entrance is not always successful. Ferton[43] records that S. subquadratus, thwarted by a guardian *malachurus* queen, dug a by-passing hole at the side, enabling it to attack the *Halictus* from the rear; she then disposed of two other guards. *Sphecodes* have also been seen attacking the host bees outside the nest.

Species mentioned

Halictus calceatus	B	Ha-lick'-tus	cal-see-ate'-us
— malachurus	B	—	mal-a-kure'-us
— zebrus	E	—	zee'-brus
— marginatus	E	—	mar-gin-ate'-us
— humeralis	A	—	hu-mer-ail'-is
— sexcinctus	E	—	sex-sink'tus
— quadricinctus	B	—	quad-ree-sink'-tus
Sphecodes subquadratus	B	Sfeck-oh'-dees	sub-quad-rate'-us
— pellucidus	B	—	pell-loo'-sid-us
— rubicundus	B	—	ruby-cund'-us

CHAPTER 19

The Osmias

THE genus *Osmia* is a very large one; its members exploit all sorts of different holes as sites for nests and use many different building materials They are middle-sized bees with long tongues, and they have a characteristic shape, shortly cylindrical and rounded fore and aft (Plate 12B). Their coats are mostly furry, in European species often brown, occasionally with black or red; a good many American species are metallic green. In contrast to the bees considered so far, they do not gather pollen on specially adapted legs but on a brush of dense hair on the underside of the abdomen.

In nearly all Osmias cells are arranged in a linear series. The habit which many species have of nesting in wood or in hollow stems has made it rather easy for a number of workers to study them by the trap-nesting technique and similar methods. Fabre[6, 9] induced the red *Osmia tricornis* to build in artificial tubes in his study. They were thus directly under his eye; but he had to leave strict instructions that the study was not to be dusted! The tubes were made of glass and covered with paper, since the bees did not like working in the light. If the tube was too wide for the bee's liking a partition of mud was built across before the cell was provisioned, but a 'dog-hole' was left near its lower side so that the bee could pass; it was eventually closed completely. A final mortar-closure when the series was complete was very solid and might take the bee a whole day to fashion. *Osmia tridentata* normally nests in bramble-stems, digging out a hole of the diameter of a lead pencil and extending to a depth of up to 20 inches. Cells are made like little barrels, ovals narrower at each end, and partitions are made of sawdust scraped from the insides of the bramble-stems. Fabre opened up the occupied bramble-stems and transferred the cells to paper-covered glass tubes, such as he had induced *O. tricornis* to nest in; he could then take these also into his study.

Krombein[13] studied particularly the wood-nesting *O. lignaria* in America. He observed that the female rested in her boring at night, her abdomen curled round with its dorsal surface just inside the entrance, so that it formed a regular plug. The ratio of pollen to nectar in the food-store was very variable. When cells were built in a wide channel, the bee had no difficulty in arranging the food as it liked. In a narrower cavity it had perforce to go in head-first to regurgitate honey from its crop, come out, turn round, then go in backwards in order to scrape the pollen off its ventral brush.

Many observations are on record concerning the nesting habits of various Osmias. A European species, *O. saundersi* (see p. 81), burrows in hard sand and lines its burrows with petals, especially those of a yellow *Cistus*.[46, 47] *O. papaveris* makes nests with single cells and uses the petals of red poppies as a lining. If these cannot be found it may use the yellow ones of the horned poppy *Glaucium*. The change may not be as great as appears to us, for honey-bees, and very probably Osmias also, cannot see red as a separate colour, so the red and yellow poppies may not appear to them so very different. A number of species nest in empty snail-shells. Ferton[43] mentions a habit, which *O. rufohirta* has, of rolling its chosen snail-shell to a suitable spot before starting to provision it. Sometimes it appeared to move its snail-shell because it was being bothered by a *Chrysis*. An *Anthidium*, to be mentioned in the next chapter, often builds its nest deep inside a shell, and an Osmia coming along later builds cells side by side in the outer part of the shell. The *Anthidium*, hatching first, has to destroy the *Osmia* cells in order to get out. Fabre,[9] scornful as ever of the doctrine of evolution by natural selection, asks why the *Osmia* hasn't invented a way of surviving this catastrophe. Some of the snail-shell dwellers use a paste of chewed-up leaves to make their partitions. *O. bicolor* does this and also may make at the entrance a sort of hut of pine needles and dried grass, while *O. aurulenta* closes the nest with a kind of felt probably derived from plants of the borage family; another species incorporates little pebbles in its plug. *Osmia rufa* (Plate 12B) and other species have made themselves troublesome by entering key-holes and building masses of cells all round the mechanism of the lock.

Fabre found that several species liked to nest in old nests of mason-bees (*Chalicodoma*; chapter 20). These were repaired and modified as necessary. The instinct to repair is, however, imperfect, for repairs are only carried out before provisioning starts. Some cells were deliberately damaged later, so that honey leaked out of them; the bee went on bringing honey for three hours making thirty-two separate journeys but never doing anything to stop the leak.

Ferton made some observations on the navigating powers of the snail-dwelling *O. rufo-hirta*.[43] He moved a half-provisioned snail-shell from its

13. Experiment on moving Osmia nest in snail-shell

original site, A. to another, B. (Fig. 13). The returning bee went repeatedly to A. before finding the shell at its new site. Ferton then removed the snail to C, a spot to the side of the line from A. to B. The Osmia, if bringing honey from any distance, went from A. to B. to C, though going directly to C. if collecting from honey close to that spot. After a further move to D, the bee tended to take a course A. C. D, having forgotten about B. Another species *O. ferruginea*, which has not got the habit of moving its snail, was much less expert than O. *rufohirta* in finding it when once it had been moved.

Fabre[9] with the artificial nests in his study and Krombein[13] with his trap-nests paid attention to the distribution of sexes in linear series of nests. In general, cells are made smaller for male bees and larger for females: also the female cells tend to be the lowest in a series, the males nearer the entrance. This is convenient as males commonly emerge first. Krombein found that with *O. lignaria* there were occasional 'mistakes', so that a female egg was laid in a 'male' cell, or, less often, the other way round. Fabre noted that in *O. tridentata*, in which the sexes are roughly equal in size, the order was much more random than in the other species, which have larger females. He was, quite correctly, convinced that the female *Osmia* could lay eggs of either sex at will: for example, when nesting in very narrow galleries or elsewhere with limited room, she would lay only male eggs. He was puzzled as to how this choice could come about; yet he dismisses as a 'new-fangled theory' the notion that the female could release sperm or withhold it and so lay a fertilized, female, or an unfertilized, male egg. Just how this is managed will be discussed later (p. 169).

Another of Fabre's interests was in how the bees in linearly arranged cells manage to emerge without harm to others in the row. He carried out experiments on the same lines as those of Cooper on *Ancistrocerus* (p. 92). As will appear, bees do not seem to have perfected a method of 'digital communication' as some wasps have done when they leave messages for their offspring in the shapes and texture of their cells. Fabre's Osmias, particularly *O. tridentata*, were normally patient and willing to wait till the emergence of those ahead of them in the queue. They made a circular hole in their ceilings as soon as they hatched and they could thus see when the coast was clear for them to proceed outwards. After a week, however, their patience tended to be exhausted and they would try to get out. They tried slipping past an unhatched cocoon, perhaps gnawing at the wall to widen the passage, but never interfering with a cocoon containing an unhatched bee. If, however, there was a blockage due to a dead bee, or a cell with an unhatched egg and mouldy provisions, they would either gnaw their way past it or, if they could, affect an exit by biting through the outside wall. When Fabre filled his tubes with alternating cells of an *Osmia* and a wasp, the bees had no compunction about forcing a way through, even crunching up the wasps on the way.

When the tubes or the cocoons in the tubes were inverted, the hatched bees managed to turn round and emerge in the right direction. If, however, the top of a channel previously open at both ends was blocked, a few only of the lower bees escaped at the bottom end. If a similar tube was placed horizontally, the bees chose the nearest exit, half going each way; but if one end was blocked, all made for the open end. Fabre supposed that the bees were sensitive to 'atmospheric influence' and this seems reasonable enough, substituting the term 'oxygen tension' for 'atmospheric influence'.

Some Osmias are particularly attached to certain flowers: different species frequent respectively poppies, thyme or convolvulus and their times of emergence correspond with the main-flowering times of their favourites. They also work at those times of day when there is the best nectar-flow, perhaps resting for a time in the heat of the day, at least in the south of France.

Towards the end of their working lives the bees' instincts seem to fail. They build defective cells, too shallow, too deep or insufficiently supplied with honey. According to Fabre *O. tricornis* 'being born to work, must die working. It sometimes continues working in order to spend its remaining strength in aimless labour.' Alternatively old bees may go berserk, starting to open and destroy cells, to eat eggs and then doing a little provisioning, egg-laying and cell-closing—all in an incomprehensible sequence.

Mention has already been made of the parasitic attacks of larvae of *Chrysis* (p. 80) and *Sapyga* (p. 87); other parasites of *Osmia* are bees of the genera *Stelis* and *Dioxys*: their larvae have aggressive sickle-shaped jaws. Others are miltogrammine flies and bee-flies (*Anthrax*; p. 141).

Species mentioned

Osmia tricornis	E	Oz'-mee-uh	try-corn'-is
— tridentata	E	—	try-dent-ate'-uh
— lignaria	A	—	lig-nair'y-uh
— saundersi	E	—	saund'ers-eye
— papaveris	E	—	pap-ay'-ver-iss
— rufohirta	E	—	roof-oh-hirt'-uh
— bicolor	B	—	by'-col-or
— aurulenta	B	—	ore-you-lent'-uh
— rufa	B	—	roof'-uh
— ferruginea	E	—	ferr-u-jinn'y-uh
Stelis	B	Ste'e-liss	
Dioxys	E	Die-ox'-iss	

CHAPTER 20

Workers in Various Crafts

To be considered in this chapter are bees of the genera *Chalicodoma, Megachile, Anthidium, Dianthidium* and *Xylocopa*; these in building their nests are respectively masons, leaf-cutters, workers in cotton and in resin, and carpenters. They are all long-tongued bees and except for *Xylocopa* collect their pollen on ventral brushes as *Osmia* does. In Britain we have only leaf-cutters and one species of *Anthidium*.

The mason-bees, *Chalicodoma*, were favourite subjects for study by Fabre.[8] The Latin name means pebble-house. He knew three species, the mason-bee of pebbles (*C. muraria*), that of the tiles (*C. pyrenaica*) and that of the shrubs (*C. rufescens*). All these bees build their nests of calcareous clay mixed with sand and saliva; this sets to form a hard mortar. The nests are rounded objects, not built in holes like those of *Osmia* and most other bees. Their outer surfaces are left rough like rustic architecture; the pebble species incorporates little stones into the outer surface. Several bees may build close together, so that their combined nests form a sort of comb and there may be co-operation between a number of bees in putting a final coat of mortar around the finished structure. *C. muraria* makes its nests under stones, balconies or on milestones: *C. pyrenaica* likes especially the under-sides of projecting roof-tiles and in such a situation there may be colonies of hundreds or even thousands, occupying several square yards: *C. rufescens* builds a much smaller, apricot-sized, nest in bushes with a twig in the middle of the structure. Since the masonry is of resistant material, old nests are often repaired and used again; or, as mentioned in the last chapter, they may be used by Osmias and other insects.

The inner walls of the nests are always smooth. They are

filled with honey, sainfoin and broom flowers being favourite sources. As was mentioned for *Osmia*, the bee commonly enters the nest head-first to disgorge its honey and then tail-first to get rid of the pollen from its brush. Fabre watched bees engaged in this exercise. When one had disgorged its honey and come out, he pushed it aside with a straw before it could carry out the tail-first procedure. The bee then went in again head-first, though it had no honey to dispose of. On emergence, it was again pushed aside and again it entered head-first; and this game could be carried on indefinitely. Evidently the tail-first entry had to follow immediately after disgorgement of honey: an interruption upset the bee by breaking the proper sequence.

Fabre calculated that collecting mud to build a cell and journeying to a sainfoin for honey involved journeying for $9\frac{1}{2}$ miles to complete and store one cell. To provision all the cells and put on the final cover would mean flying for 275 miles.

The habit of *C. muraria* of building on large pebbles made it possible to move the position of the nests and observe the results. Moving the pebble for only one yard upset the bee; she found and inspected the nest but didn't carry on with work upon it. If, however, a second pebble plus nest was placed on the original site, she would proceed to finish or provision that, even though the pebble looked quite unlike the original one. Fabre kept on interchanging two near-by nests with similar results. If a bee in the middle of masonic work was given a cell already full of honey, she nevertheless went on with quite unnecessary building; and if a forager was provided with a nest already full of honey she would put her honey into an incompletely built cell or open a closed full cell, deposit her honey and lay an egg in that. Fabre also carried out experiments on the homing powers of mason-bees, with results similar to those recorded for *Polistes* (p. 98), and other species: most bees readily found their way home from $2\frac{1}{2}$ miles away.

When young bees hatch out from the pupa, they have of

course to make their way through the masonry of the nest wall and this they do without difficulty: masonry is something with which instinct equips them to cope. If, however, the nest is placed under a funnel covered with gauze, the bee which has found its way through the rock-like covering to the nest is completely baffled by the layer of gauze, Fabre covered a nest with brown paper. If this was applied closely to the nest, the bee got out without difficulty, taking the paper-cover in its stride. But if there was a one centimetre gap between nest and paper, the bee failed to get past it.

Besides cuckoo-bees of the genera *Stelis* and *Dioxys*, mason-bees are greatly troubled by two other parasites. A black-and-yellow parasitic hymenopteron, *Leucospis*, has a very slender rapier-like ovipositor, as thin as a horse-hair. The parasite somehow locates the position of the cell under its covering of masonry and manages to insinuate its very slender weapon through the solid nest wall and to lay an egg on the bee-larva. She may do this a number of times and even inadvertently deposit several eggs in one cell. When this happens, only one parasite survives, as was the case with *Chrysis* larvae (p. 80). Usually the *Leucospis* feeds on fully-fed mason-bee larvae. sucking them dry.

The striking black-and-white bee-fly *Anthrax* achieves the same end in another way. The fly, like its close relative *Bombylius*, hovers in front of the nest-entrance as it deposits an egg. From this hatches what Fabre[7] calls the primary larva, an exceedingly minute eel-like creature which moves much as a looper-caterpillar does. This manages to creep into the mason-bee's nest, finding, we presume, minute crevices as the ovipositor of the *Leucospis* had done. It may, however, take weeks or even months before it manages to get in. Once inside, it waits patiently if necessary until the bee larva has finished feeding. It then moults and turns into a secondary larva of very different appearance. This is relatively immobile with a humped thorax and a nipple-like head with the opening of its gullet at the bottom of a little funnel: there are no teeth.

The opening is likened by Fabre to 'a sucker or cupping glass. Its attack is a mere kiss, but what a perfidious kiss. . . . It does not eat, it inhales.' When an *Anthrax* larva had finished feeding Fabre took the residue of the bee-larva, a mere granule, and blew it up with a fine tube, restoring the larval shape: the skin of the larva was intact. This technique is made possible because the attack is made at the stage of 'histolysis', when all the larval tissues are dissolved preparatory to being re-assembled into the form of an adult bee.

To get out of the masonry, the *Anthrax* pupa is finished with 'a diadem of spikes on its head, a many-bladed ploughshare behind, four climbing belts or graters on its back' and is covered all over with long stiff backward-pointing bristles. When the time for emergence comes, it arches itself for a blow, then releases itself, striking the plug ahead of it with its barbed forehead. It then sways and fidgets, grinding away at the obstacle. The stiff hairs and apparatus on its back help to anchor it while the frontal assault is in progress. Finally the head and thorax of the fly pupa project from a hole and from this the fly comes forth.

Another insect to be found in *Chalicodoma* nests is the little parasitic hymenopteron, *Monodontomerus*.[7] This is a parasite not only of *Chalicodoma* but also, and perhaps preferentially, of *Stelis*, the cuckoo-bee which is itself preying on the mason-bee: this is an example of hyperparasitism. Many parasite larvae consume that of the host, any number from four to thirty-six, the number being roughly related to the size of the host. Eggs are inserted by the same method as adopted by *Leucospis*, and the larva is sucked dry by the *Anthrax*'s kiss-technique. When the time for emergence comes the *Monodontomerus* take turns at excavating a hole; one works away and others stand by awaiting their turn for labour. Of nine cells in one of Fabre's *C. muraria* nests, three were occupied by *Anthrax*, two by *Leucospis*, two by *Stelis*, one by *Chalcids* (another kind of hymenopterous parasite) and only one by a mason-bee.

The leaf-cutter bees, *Megachile*, are familiar to most people through their habit of cutting out circular discs or ovals from the leaves of rose-trees and other shrubs. The larger species are not unlike honey-bees, but with a rather square-ended abdomen; they are easily distinguished from them if a glimpse is caught of the pollen-laden under surface of the abdomen. Our main interest lies in their nest-building habits. Nests are built in all sorts of holes, often bored out in wood, less often made in the ground. Used holes of other species are often used, the material of a previous occupant being first cleared out.

The cells are lined and closed with pieces of leaf. The bee seems to be using her own body as a measuring instrument: yet the leaf-pieces are not always of the same size and shape. *Megachile brevis* and other species make the inner ends of cells and cell-walls of rectangular overlapping cuttings, bent inward at the inner end to form the base of a cup-shaped cell. The edges are 'mouthed' and so made to stick together. Fifteen rectangular bits were used to make the cell; after provisioning and egg-laying, the top was sealed with circular pieces, and the process was then repeated. The whole sequence could be lifted out as a cylinder. On the other hand, Fabre[9] found with a species he studied that removal of the newly-provisioned cells made them fall to pieces; only when the full-grown larvae had consolidated the structure with silk did it hold together. Some species make their plugs partly of loosely-shingled pieces but with tightly packed circles at the entrance. *M. policaris* makes either single cells or brood-cells containing several larvae.[13] In the latter case the female stores a pollen-nectar mixture, makes a hollow and oviposits in it, then adds more food—and so on. She does not line her cells but puts gummy leaf-pulp at the bottom, with a final closure of leaflets of various plants. There may or may not be partitions. Presence of several larvae in one cell seems not to lead to cannibalism. A few Megachiles make a masonry lid on top of the closing discs of leaves.[48] Fabre found a leaf-cutter

(*M. albocincta*) placing strong rough leaves at the bottom of its burrow but using only smooth ones for lining. On the whole leaves are chosen not because they come from a particular plant, but because they are of the right texture and pliability. Obviously, if a bee finds a bush with suitable leaves it will keep returning to the same source. In the construction of the closing plug, the first pieces were of exactly the right size for the diameter of the burrow, later ones rather larger so that they had to be pressed down and so were made slightly concave. A nest of one species contained 1,064 pieces of leaf, but such a large number is unusual. A few species use flower-petals instead of leaves. As was recorded in the case of *Osmia*, the sequence of operations may go quite wrong, probably when bees are at the end of their lives: Fabre found galleries completely filled with rough pieces of leaf, no cells being formed, no provisions stored nor eggs laid.

The special parasites of *Megachile* belong to the genus *Coelioxys*. These are rather shiny black bees with some small white bands of down at the sides of the abdomen, which is unusual among bees in tapering to a point. (Plate 14B). Like other cuckoos they enter and lay in their hosts' nests; they are said to plunge their egg into the middle of a honey-pollen mass where it will remain undetected.

Members of the genus *Anthidium* are workers in cotton.[9] They make their nests in old earthworm holes in the ground, in borings in wood and elsewhere. The 'cotton' is obtained by scraping the leaves or stems of any downy plant. (Plate 14A.) *A. florentinum* uses the pappi on various composite seeds—the feathery parachutes which help to disperse seeds in the wind. The 'cotton' brought home is teased, torn apart, loosened and made into a sort of felt for lining the nest. A cylinder of felt can be extracted from the nest, but the partitioning into cells cannot be detected until the larvae have made their cocoons, when knots between the individual cells become visible. A coarser flock may be used for closing the cell than for the lining. *A. manicatum*, the only species occurring in

11(A) *Andrena flavipes*

11(B) *Nomada fucata* investigating an *Andrena* burrow

12(A) Male *Anthophora acervorum* (notice hairs on middle legs)

12(B) *Osmia rufa*
approaching a flower

Britain, uses burrows in the ground about 10 cm deep. Ferton records that there may be, besides the felt, a barricade 3 cm down made of little white stones each the size of a small pin's head. He timed one bee making twenty-nine trips with material into its nest within two minutes. Krombein found *A. maculosum*, nesting in wood, sealing its nest with matted fibres, sometimes interpolating a section of pebbles.

The food store consists of a rather liquid honey on top of which the egg floats. The larva begins to void waste matter when half grown. This, in the form of pin-head sized granules, is stuck on to the circumference of the cell, finally making a screen all round the larva. When it subsequently makes its cocoon it collects these granules and uses them to strengthen its wall.

In contrast to other bees, the males of *Anthidium* are larger than the female. Those of *A. manicatum* are very aggressive and actively chase any hymenoptera of about their own size, or even drone-flies, presumably hoping they will turn out to be females of their own species. They used to collect the 'cotton' from the spider-web house-leeks (*Sempervivum arachnoideum*) on my father's rockery and, as a boy, I was fascinated by the aggressive tactics of the males.

Those species which use resins instead of cotton are now placed in a separate genus, *Dianthidium*. Fabre studied several species, two of which nested in old snail shells. Often the shells were buried under heaps of stones. Because the male bees are larger than the females, their cells, too, are larger: so in snail shells a cell for the female was deeper where the passage-way was narrower. Mention has already been made of the troubles which occur when an *Osmia* nests on top of *Anthidium* cells in a snail. The resin to line the nests is obtained from juniper and possibly other conifers. Other *Anthidium* species build their nests wholly of resin and these nests are fixed under stones, forming little balls the size of a fist or an apple, brown, hard and slightly sticky. They may be decorated with all sorts of oddments including ants'-heads. Their nests are, however, no more proof against penetration by *Anthrax* larvae than are those of the mason-bee.

The carpenter-bees, *Xylocopa*, are very different from the other craft-workers in this chapter. There are rather similar species on both sides of the Atlantic, though the European *X. violacea* does not reach Britain. They resemble large jet-black bumble-bees with violet-black wings. They nest in tunnels in wood, preferring old borings either in sound or dead wood, and according to Krombein[13] colonies of the American *X. virginica* have used the same tunnels for fourteen years. In Florida, nesting goes on throughout the year. Wood-pulp is used for partitions between cells: this the female 'winds spirally, ring within ring, until the intercellular space is sealed'. Females may stand guard outside their nests instead of making an impermeable closure at the entrance.

Species mentioned

Chalicodoma muraria	E	Calico-doh'-muh mu-rare'-ee-uh	
— pyrenaica	E	— pirr-en-ay'-i-kuh	
— rufescens	E	— roof-ess'-enz	
Megachile brevis	E	Meg-a-kile'-ee brevv'is	
— policaris	A	— pol-ee-care'-is	
— albocincta	E	— al-boh-sink'-tuh	
Coelioxys rufescens	B	Seely-ox'-iss roo-fess'-ens	
Anthidium florentinum	E	An-thid'-ee-um flo-rent-ine'-um	
— manicatum	B	— manny-kate'-um	
— maculosum	E	— mack-u-low'-sum	
Dianthidium	E	Die-an-thid'-ee-um	
Xylocopa violacea	E	Zy-loh-cope'-uh vi-o-lace'-ee-uh	
— virginica	A	— virgin'-ick-uh	

Parasites

Leucospis		Lu-koss'-piss
Monodontomerus		Monno-donnto-meer'-us

CHAPTER 21

Bumble-bees: Life History

THE genus *Bombus* includes many species of bumble-bees. Pedants have urged that they should properly be called Humble-bees, but this name has almost died out. Bombus is a genus extending into northern latitudes, even up to the arctic, though there are tropical species. In Britain there are twenty-five species. In the short sub-arctic summer the *Bombi* can only work for a few weeks in the year; on the other hand tropical species may nest and work all round the calendar. Their social life has developed along the lines described for the Halicti, but has gone a good deal further – not, however, as far as that of the honey-bees (*Apis*). Some aspects of their behaviour will only be touched on briefly as they will be more fully dealt with in the four chapters on honey-bees.

What we know of their habits is described in two very informative books, one, by F. W. L. Sladen,[23] published in 1912, the other, by J. B. Free and C. G. Butler,[12] in 1959. Both these works describe how bumble-bees can be induced to build in observation nests, so designed that activities within the nest can be kept under constant observation.

As with the *Halicti* and social wasps, the females are fertilized in the autumn, hibernate and start to nest in the spring. On warm days in early spring they may come out and sun themselves, then retreat into their holes underground and become torpid once more. It is when their hitherto thread-like ovaries have begun to develop that they begin to look for nesting places. Old mouse nests are favourite sites. Some species nest at the end of tunnels several feet long and others only a few inches down. Old bird's nests and thatched roofs may also be used. Many more nests will be found in earth-banks, commons and rough grass than in intensively cultivated land.

To begin with, the nest is only an inch or so across and lined with grass. The queen stays inside for a few days to dry it off, then secretes wax to make a cup-shaped cell. The first nectar collected is simply deposited in the nest material, but when the cell is ready, nectar and pollen are placed in it and eggs are laid, eight to twelve in one cell. The cell is closed and the queen then builds and fills a honey-pot near the cell-entrance, half-an-inch across, three-quarters-of-an-inch deep. From the honey in this the queen can sustain herself in the bad weather which too often interrupts her work early in the year. When the eggs hatch and the larvae begin to feed, the cell is opened so that more food can be supplied; the cell is also enlarged as necessary. Each larva when full-fed spins its own cocoon and when all have done so, the wax is removed for building fresh cells on top of the cocoons. The queen incubates her brood, coming to lie in a groove between the cells, covering as many as possible with her body. The bees produced are at first all workers, infertile females. Time from egg-laying to hatching out of the bees is around twenty-four to forty-four days, but it varies with different species and according to weather and food-supply. As soon as the young bees are out and dried off, they begin to forage, and the queen then ceases to do so. When the second batch of eggs hatches, the young workers will help the queen to care for the larvae. The queen never becomes a mere egg-laying machine like a queen honey-bee. She will, however, increase her production as the season advances, laying up to a batch of eggs every day.

The size of the comb gradually increases, (Plate 15A) but there are no more larvae produced than the workers can care for. Early in the colony's life, some ripe eggs in the queen's ovaries are re-absorbed, thus avoiding excessive production of larvae. There may also be some cannibalism. The queen's old honey-pot is no longer used for storing honey; instead the workers store honey in empty cocoons. Some species also store pollen in cylinders which may be as much as six inches high. According to F. W. L. Sladen bumble-bees are of two sorts, pocket makers and pollen storers. The former make pockets filled with pollen placed outside groups of developing larvae; these feed directly on this pollen. The latter feed the larvae on regurgitated pollen and nectar. Some of the early workers are hardly bigger than a house fly; size probably depends on how favourably the cell is placed for being provisioned.

Nests increase in size by addition of fresh material, grass and moss, which gets woven in; they may ultimately be three to nine inches across. A canopy of wax may be built over the combs. Through the early summer more and more workers are produced; there may be three hundred to four hundred in nests of some species, while others rarely have more than one hundred. Finally production of new queens and males begins. There is no evidence that special food, as with *Apis*, makes queens instead of workers; it seems rather to be a matter of quantities of food. The larvae get more food when there are lots of foragers bringing it in and it appears that queens begin to be reared when the worker: larva ratio is 1:1. This comes about in part because the ageing queen comes to lay fewer eggs at a time when there are plentiful workers. It is, however, a mystery why this state of affairs also leads to production of males, which are reared in the proportion of about two males to each young queen. Up in the arctic very few workers can be reared before queen production has to begin.

The young queens at first help in foraging and other duties. The males or drones, however, leave the nest and they may or may not return. Males of some species hover round the outside of a nest awaiting the appearance of virgin queens, and these may even be pursued into the nest. Others—and these are species of which the males have very big eyes— station themselves on a spot with a good view, and chase after anything which might be a young queen. Yet others 'fly along established routes in a seemingly endless procession', stopping to hover round particular places which are apparently marked

early in the morning with a specific scent. One *B. terrestris**
made thirty-five circuits of a three hundred yard course in an
hour and a half, visiting on the way twenty-seven special
stopping places. The drones are attracted by the scent of
females: even a dead queen on a path was a great source of
interest. The males and workers all die off in due course:
those of *B. pratorum*, the 'early-nesting bumble-bee' do so by
July, others not till September. The young queens on the
other hand fill up their honey-stomachs with food and seek
winter-quarters; these are often in banks facing north or north-
west, a position which ensures that the spring sun will not
awaken them too soon.

Division of labour among the workers is less evident than
with honey-bees; on the whole the smaller bees tend to act
as 'house-bees', the larger ones as foragers. This is obviously
advantageous: smaller ones are more mobile about the nest
and larger ones can bring in more food. Free found that most
bees in a colony kept to one kind of duty, but about a third
of them were less consistent. If enough foragers are caught and
removed, house-bees turn to foraging. Worker bees of the
common red-brown *B. agrorum* were found to live less than
three weeks on the average; other species survived considerably
longer.

As with other social bees and wasps, workers may lay eggs,
especially in the absence of a queen. If several workers are
kept in an artificial nest, the one with the most ovarian development attains a state of dominance over the others. Then there
may be a No. 2 who lords it over all but No. 1 and so on: a
peck-order such as was described for *Polistes* (p. 97). It is
mainly No. 1 which lays eggs—all destined, of course, to
yield only males. The order of dominance may change. If a
No. 1 is introduced into another queen-less colony she will
become dominant unless there is already a forceful No. 1 in
possession, in which case she will usually be worsted. An

---

* The species commonly so called; there is dispute as to the correctness of
the name.

introduced queen will usually take over the No. 1 position, though perhaps only after a fight. There is no evidence among *Bombi* of a queen-substance, such as will be described in chapter 24 for *Apis*. Presence of the queen does, however, seem to inhibit ovary-development and egg-laying by the workers. The workers may, when production of new queens and males begins, start to eat eggs and even open cells for this purpose, though the queen does her best to thwart this. It happens especially with *B. terrestris, lucorum* and *lapidarius,* all pollen storers.

Bombi sometimes raid other nests for honey or even hives of honey-bees but the latter may raid in their turn: so may wasps. Raiders sometimes carry off their booty unmolested, but they may be attacked and killed especially if of a different species. An attacked *Bombus* assumes a characteristic defensive attitude, lying on her back with the middle and hind legs raised; she buzzes at the same time. Recognition of a nest's own bees seems to depend largely on odour acquired from the nest, rather than on smell of a particular kind of food being brought in. Workers confined for a few hours in a strange nest and then brought home were attacked as if they were strangers. Only for a few strong colonies of *B. terrestris* and *lucorum* are there special guards posted at the nest's entrance.

A bumble-bee's sting can be fatal to another *Bombus* or other bees. Some people consider it more painful than a honey-bee's sting, others less so. Those who are allergic to *Apis* stings are as a rule not affected unusually by *Bombus* stings.

Species mentioned

Bombus terrestris	B	Bom'-bus terr-est'-ris
— pratorum	B	— pray-tor'-um
— lapidarius	B	— lappy-dare'-ee-us
— humilis	B	— hew'-mill-iss

Species mentioned (*cont.*)

| Bombus lucorum | B | — | lucore'-um |
| — agrorum | B | — | agg-roar'-um |

CHAPTER 22

Bumble-bees: Flowers and Enemies

BUMBLE-BEE foragers collect their pollen on a brush inside the basitarsus of the hind leg (see Fig. 14). Thence it is scraped off by a comb on the end of the tibia into a receiver. When the leg is straightened, a projection on the metatarsus compresses it and pushes it into the lower end of the pollen-basket where there is a break in the wall of hairs, permitting its entry. The metatarsus of the middle leg now comes into play and pats it down. The pollen-mass is made gradually to slide up the smooth floor of the pollen-basket till that is sufficiently full. A full pollen-load may come to as much as sixty per cent of the bee's weight. A bumble-bee usually confines itself on any one day to collecting either nectar or pollen, though pollen-gatherers usually bring home a little nectar also. The relative amounts of pollen and

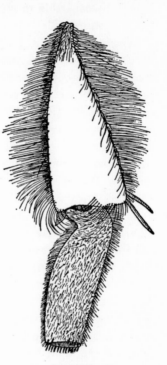

14. Leg of Bombus

nectar collected depend upon the colony's needs at the time. If nectar or pollen supplements are deliberately added to an experimental nest, the bees will bring back more of the other food to redress the balance.

Bombi will go foraging in bad weather when most other

insects, even honey-bees, remain in shelter. Early in the year, when reserves are poor, or the queen is on her own, work may be carried on till an hour after sunset. In the arctic summer, bees may forage all night long, and house-bees anywhere may be busy about their duties all through the night. Foragers may be overtaken by bad weather or darkness and may sleep out, and then workers may be found in hollyhock and other flowers in the early morning or in bad weather, almost torpid, though perhaps able to give the defence-reaction of raising two legs on one side.

Experiments will be described in chapter 26 on the attractiveness of colours and scents to honey-bees. Similar experiments, carried out on *Bombi*, will not be described in detail, as the results are much the same as for *Apis*. Bumble-bees, like honey-bees, are red-blind. The sizes and shapes of petals are of importance in attracting them, irregularly shaped flowers such as monkshood (*Aconitum*) and snapdragon (*Antirrhinum*) being favourites. However, there are less conspicuous flowers such as those of sallow and *Cotoneaster* which are much visited. Different species have their preferences, those with longer tongues visiting especially flowers with tubular corollas. Many species are greatly attracted to clover and these are of great economic importance, as a farmer is greatly concerned about the yield of seed from his clover-crop. Darwin's observations on this matter are well known. He found that one hundred heads of red clover visited by bees yielded 2,700 seeds, while those protected against bee-visits yielded none. It is the long-tongued *Bombi* which are important, not the short-tongued ones nor yet the honey-bees. Flowers which are robbed by having the tubular corolla bitten through near the base will not set much seed. Darwin suggested that the yield of clover-seed depended on the local abundance of cats, for these caught the mice which destroyed the nests of the bumble-bees which were essential for cross-pollinating the clover. Huxley added old maids to the complex ecological picture, since these tended to have many cats. The example is an interesting one, though the accuracy of the observation is rather in question: for one thing the mice are not wholly harmful to the bees, for their abandoned nests provide useful sites for *Bombus* homes.

Bumble-bees are also more useful pollinators of fruit trees than are honey-bees. They work longer hours and in unfavourable weather. More important still, they wander more and are therefore more likely to pass from tree to tree, an important matter in the case of apple and other trees which are not self-fertile. Tinbergen[25] noticed that marked *Bombi* when visiting hound's-tongue (*Cynoglossum*) flew, where plants were rather far apart, in a regular order from one to another. If one was dug up and removed, the bees still paid a routine call at the spot where it used to grow. The bees also, when leaving a profitable clump of hound's-tongue, circled round for a locality-study before returning to the nest. Hound's-tongue flowers are not very conspicuous: where more conspicuous flowers such as fox-gloves were concerned, no such locality-study was made; the plants could doubtless be seen from a distance. Free and Butler suggest that, since *Bombi* are so useful for the pollination of crops, more should be done to encourage them. Ploughing an odd bit of ground and scattering hay over it, or leaving areas of rough grass should provide nesting sites. Introduction of bumble-bees into New Zealand made all the difference to seed-setting in clover crops.

The subject of the parasites of bees is by no means exhausted, for new aspects are presented by those attacking bumble-bees. Chief among them are the cuckoo-bumble-bees of the genus *Psithyrus*, a name suggesting a whispering sound which they are said to make. To the uninitiated they look like other bumble-bees, most being black with yellow or white bands; one, parasitic on the black red-tailed bumble-bee, *B. lapidarius*, is coloured much like its host. Female *Psithyri* do not have pollen-collecting apparatus on their hind legs like Bombi (Plate 15B); further, like many other parasites, they have hard horny coats of mail, not easily penetrated by hostile stings. Richards[57] lists twenty-seven characters distinguishing *Psithyri* from *Bombi*; nearly all these are related to their parasitic habits. They have pointed mandibles, useful for attack, and besides the horny coats there are other devices to defend against an enemy's stings: there is hardly any gap between head and thorax, the plates covering the abdomen overlap and there is a box-like closure of the anal orifice.

Queen *Psithyri*, like their hosts, are fertilized in the autumn, hibernate and begin their activities in the spring. This, however, is a good deal later than for *Bombi*: they only take steps to invade a nest where there are enough workers to provide for bringing up the parasites' offspring. The nest-hunting female is probably guided largely by scent. When she enters a *Bombus* nest she may at first hide beneath the comb and nest-material; this probably helps her to acquire the nest-odour so that she may escape from being attacked. When she comes forth she may be tolerated, but will probably elicit a hostile reaction from the host-queen when she begins to lay. Nests may, however, contain queens of host and parasite living in apparent harmony; in such a nest, however, few if any *Bombus* larvae will be reared. To Sladen, the *Bombus* queen in such cases seems despondent and apathetic. The *Psithyrus* may eat Bombus eggs and even larvae. As to her other activities, one has been seen to lay twenty-three eggs in six minutes. She never does anything to help feed or otherwise attend to her own brood, leaving it all to the *Bombus* workers; at most she guards her eggs for a while.

Events may, however, take a different course: the cuckoo queen may be attacked as soon as she enters. Workers may 'ball' her, swarming all over her, each trying to sting; and she may well be killed in the end, perhaps after slaying a number of the defenders. The fight may, however, be between the two queens. According to Sladen queens of *Ps. rupestris* and *vestalis* always fight and kill their respective host queen, *B. lapidarius* and *terrestris*. Against too vigorous an onslaught a *Psithyrus* queen can draw up her legs and remain immobile, defying her enemies to penetrate her strong defence. When the new brood of *Psithyri* hatch out, the young queens tend to remain for some time in the nest, while the males go forth and can be seen feeding or drowsing on flowers: they also fly around circular 'beats' as described for the male *Bombi*, on the look-out for mates.

It is not hard to see how the parasitic habit arose, for the tendency is already present among the *Bombi* themselves. Queens of *B. terrestris* and *lapidarius* are especially likely to try to take over an existing colony rather than founding one for themselves. The queen *B. terrestris* even has a stronger more curved sting than other *Bombi*, resembling that of *Psithyrus*. The queen in possession often resists this invasion successfully: Sladen found twenty dead *terrestris* queens outside one nest. Occasionally a queen may take over the nest of another species: *B. terrestris* may thus be found as the occupying queen of a *B. lucorum* nest, and in America, *B. affinis* may take over from *B. terricola*. There is little difference between this and the behaviour of the *Psithyrus* queens.

Another interesting question concerns the origin of the genus *Psithyrus*. Where we find species associated with different *Bombi* and often, though not invariably, resembling them in colour-pattern, we are tempted to believe that each *Psithyrus* may have arisen from its corresponding *Bombus*, some members of this genus having specialized in parasitism on the lines described in the preceding paragraph. On the other hand, when we consider Richards' twenty-seven characters separating all *Bombi* from all *Psithyri* we may well believe that the two genera diverged a very long time ago, and that species have differentiated subsequently, each evolving in parallel with its host. Richards[57] discusses fully the pros and cons of each view-point. He suggests that the parasitic habit may have evolved at the northern edge of the range of more southerly species. These would tend to come out of hibernation later in the year than northerners and might thus find it hard to discover unoccupied nesting sites and so be tempted to take advantage of the presence of already active colonies of the northerners. He comes down in the end rather on the side of multiple origins, supposing that the twenty-seven points of similarity are there because of 'convergent evolution'.

These cuckoos are by no means the only enemies of bumble-bees. Badgers dig out their nests in Europe and skunks in America; field-mice also destroy many. More than sixty species of beetles are recorded from their nests. The larger species of robber-fly catch and devour the adult bees. Flies of the parasitic family *Conopidae* attack the bees in mid-air and oviposit on the membranes between their abdominal segments. The fly larvae go on developing in the bee, which remains active until the larva is full-grown, when it pupates inside the bee's abdomen. There are miltogrammine flies of the genus *Brachycoma*, rather larger than those attacking solitary bees and wasps. The flies lay just-hatched larvae in the bees' nests, but there is no attack on the bee-larvae until these are full-fed and spinning up. There may be several generations of flies a year. Most nests in Britain are affected and some may be wholly destroyed. Larvae of the wax-moth, *Aphomia sociella*, better known as a pest of honey-bee hives, make silken tunnels in the wax, eating not only it but the bee larvae and pupae, too.

The last enemy is a worm, *Sphaerularia bombi*. These tiny nematode worms live free in the soil during the summer. In the autumn fertilized females penetrate the abdomens of hibernating queen bumble-bees. At the end of the period of hibernation, the genital organs of the female worm develop a sac which is soon full of eggs. This sac reaches a volume 20,000 times the size of the rest of the worm, which can be seen as a mere tiny thread-like appendage at the end of the relatively huge sac. In due course the sac ruptures and liberates the eggs, and the hatching larvae can roam freely. Queen *Bombi* may be infested by many worms and in them the ovaries never develop so that they have no inclination to nest. When the worm larvae are liberated, the queens soon die.

There are also larvae of flies and other insects which are useful scavengers. A large hover-fly, *Volucella bombylans*, is a common one, existing in two colour-forms, either black-and-yellow-banded or black with a red tail, colours corresponding to those of common European *Bombi*.

Such a list of the enemies of bumble-bees must make one marvel that this is, nevertheless, one of the most successful genera of bees we know.

Species mentioned

—	Bombus terricola	A	Bom'-bus terr-ick'-oh-luh
—	affinis	A	— aff-ine'-is
Psithyrus vestalis		B	Psith'-ee-rus vest-ail'-is
—	rupestris	B	— roo-pest'-ris

Parasites

Conopidae	Con-ope'-idee
Brachycoma	Bracky-com'a
Aphomia sociella	Aff-oh'-me-uh so-she-ell'a
Sphaerularia bombi	Sfere-u-lairy'-uh bom'-bi
Volucella bombylans	Voll-u-sell'-uh bomb'y-lans

CHAPTER 23

The Honey-bee: Origins and Senses

WE NOW come to the insects in which evolution has led to the most beautiful and complex social organization to be seen among all the wasps and bees. We will not now make invidious comparisons with the ants nor with the termites; in both of these groups social evolution has also reached a high state of complexity. A great deal is known about honey-bees, members of the genus *Apis*, because of their importance to man as a source of honey.

Two excellent books on the lives of honey-bees are available; these are *The World of the Honey-Bee* by Colin Butler[1] and *The Behaviour and Social Life of Honey-Bees* by C. R. Ribbands.[19] I have drawn freely for the next four chapters on their writing but can, of course, give the facts only in outline. Butler's book gives a more 'popular' account of the subject: that of Ribbands gives more fully documented details.

Four species of the genus *Apis* are known; three have their original home in southern Asia: the fourth, our *Apis mellifera*, inhabits Europe and parts of Africa. The four species show gradations in their degree of social development. The giant honey-bee, *Apis dorsata*, has smoky wings and builds combs in the open air beneath limbs of forest trees, overhanging rocks and in similar places; these nests may be up to 5-6 feet long. Because of this building habit it has not been possible for man to persuade these bees, for his own convenience, to nest in hives or other shelters near his home. The same applies to the little honey-bee, *Apis florea*: the open-air nests of this species are quite small and suspended from the branches of trees or shrubs. Because they inhabit tropical countries they have not needed, as *A. mellifera* has, to make such regular provision for storing food and keeping warm

13(A) *Halictus sexcinctus* queen: the portress

13(B) *Halictus calceatus* male on a Buddleia

14(A) *Anthidium manicatum* collecting a ball of cotton

14(B) The cuckoo-bee *Coelioxys rufescens*

their nests through a cold season. They have a habit, not seen in our honey-bee, of 'absconding' when necessary; that is, the entire colony, queen and all, deserts its nest when things are unfavourable and founds a new colony elsewhere. It may be done when there is bad infestation by wax-moths or after invasion by ants or termites. It is something different from 'swarming', to be considered later, which is a means by which colonies are divided into two or more parts.

The fourth member of the genus, the subspecies *indica* of *Apis cerana*, is very like *A. mellifera*; it, too, nests naturally in hollow trees or other cavities and, like it, can therefore be readily persuaded to occupy hives. It is the 'domestic' bee of the East. Butler points out, however, that bees have never really been domesticated; they are as wild as wild bees in their habits; man has merely persuaded them to occupy lodgings from which he can readily steal their honey. He has, however, had some success in breeding superior races of bee.

A. indica retains rather more primitive characteristics than does *mellifera*. Colonies may 'abscond' as do those of the giant and little bees and like those species the workers have a defence mechanism for warding off the attacks of enemies such as ants, hornets and wax-moths: they shake their bodies in concert from side to side with a 'shimmering' movement, a manœuvre which seems to lead to the desired result.

A. mellifera is known in a number of races or sub-species. The African bees differ from the European and the latter are again divisible. A dark black-brown bee (race *lehzeni*) was formerly the prevalent one in Britain. In 1905 there began an epidemic called Isle of Wight disease from its place of origin: this was due to various factors including infestation with a parasitic mite. It spread all over Britain, killing 90% of the bees. Re-stocking was carried out mainly with bees of the Italian strain, var. *ligustica*; workers of this have yellow abdominal bands. Other races are the typical *mellifera*, the Carniolian bee (*carnica*) and the Caucasian (*caucasica*). The

Italian bees have been distributed widely, also, in America, Australia and New Zealand.

As with the social wasps and the bumble bees, a colony consists of queen, males or drones, and imperfect females or workers. In contrast to other groups, at least as they are known in temperate climes, the colonies are perennial: they live through the winter and do not have to be started afresh each spring.

In describing the life of the honey-bee it will be convenient, first, to describe what is known of its senses, so that we can see how the world looks to a bee; then to consider all aspects of reproduction; then the various activities going on in the hive ancillary to that central function; and finally bees in relation to flowers and food-gatherings. All these matters are of course intricately linked with each other.

The eyes of bees, as of other insects, consist of some thousands of small facets; these are the ends of radially placed tubes with a sensitive area at the bottom. Each tube points in a slightly different direction; the sensation conveyed by the light reaching each sensitive spot is homogeneous. The bee therefore sees its surroundings as a mosaic, rather as we see things on a television screen. There is however much less detail, for the worker bees' eyes have only 6,300 facets, compared with vastly larger numbers of dots on the TV screen. As the foraging bee is constantly moving, the images it sees keep changing and giving a flickering effect. Bees seem very sensitive to such flicker, for they are more attracted to flowers gently moving in the wind than to those which are stationary.

There has been much work on the sensitivity of bees to patterns of various kinds. In these experiments and in those concerning colour-perception the general plan is to offer to the bees saucers of sugar placed on backgrounds of different colours or patterns. Some saucers will only contain water and the bees are readily trained to associate a particular colour or pattern with presence of food. Then the guiding signs can be varied and information thus gathered as to the bees' preferences and their ability to distinguish minor differences. Bees could not, it turned out, tell the difference between solid figures such as circles, squares and triangles, but were more affected by complex patterns, preferring complexity to simplicity.

Even more interesting are the results of tests on colour perception. Bees can definitely distinguish between the colours orange, yellow, green, violet and purple, and control experiments show that this is not merely an appreciation of different degrees of brightness. However, they apparently cannot perceive red, persistently confusing this with black or dark grey. To make up for this they are able to see ultraviolet, as human beings cannot. Scarlet poppies, which attract bees, do so not because they are red but because they reflect much ultraviolet and may appear to them to be bluish. White objects may or may not reflect ultraviolet rays and so things which are whiter than white to our eyes may be either white or blue-green to a bee. It seems likely that bees perceive only a limited range of colour differences. As we shall see in a later chapter, (p. 181) bees have power of perceiving the plane of polarized light, a function lacking in man—unless he uses specially designed scientific instruments.

Bees have organs sensitive to taste in their antennae and also in their legs, especially their front legs. As would be expected, they perceive and appreciate sweetness, but some substances such as saccharin which taste sweet to us convey no such sensation to bees. They prefer sucrose to glucose and glucose to fructose; and are particularly attracted to a mixture of these sugars in equal parts; these substances do in fact make up the greater part of honey. Bees do not like bitter or acid tastes and only tolerate minimal quantities of salt.

The antennae are also the site for detection of smells. Bees will not respond to scents after amputation of antennae; the eight terminal segments are the important ones. The matter has been investigated by methods similar to those used in studying the vision of bees. Von Frisch offered his bees boxes scented with various essential oils; one box was scented with oil from sweet Italian orange. The bees had been previously trained to associate this odour, the 'training scent', with presence of sugar. They visited the boxes with training scent much more often than the others but could not distinguish it from the odours of two other citrus fruits. Ribbands found that bees could perceive scents too weak to be detected by man, their thresholds for perception being ten to a hundred times lower than it is for us.

There will be more to say later about flower-scents in relation to foraging. Scent is also important in permitting bees to distinguish between their own hive-mates and strangers. Differences between the scents of bees seem to depend partly on inborn factors and partly on the food the bees have recently taken. A colony was deprived of its queen and divided into three, two being fed on heather honey, one on syrup. After a week they had acquired distinctive body-odours. The two heather-fed sections mixed happily together, but there was mutual suspicion and no free mixing occurred between the heather bees and the syrup-fed ones. Bees have scent-glands which they can expose at will and scent liberated in this way may help to attract the bees' companions to a tasty dish. Though queen bees make piping sounds and workers moaning ones, there is dispute as to whether bees hear. Probably they do not hear as we do, but may respond to certain vibrations.

A sense of touch is, however, very necessary because they spend much of their lives in complete darkness within their hives. By means of a tactile sense they quickly learn their way through a maze to a source of syrup; they do this more readily if the floor is made of different materials—tin, glass, cardboard and so on. Curiously, quite an independent course of learning is necessary for getting back home again from the syrup; bees cannot use what they learnt on the way out to help them on the way back in. The sense of touch is located partly in the antennae, but probably also elsewhere. We have been able to learn about such things partly because of the convenient fact, mentioned earlier, that bees cannot perceive red; so observations can be made in red light, which as far as the bees are concerned, represents total darkness.

Bees have yet other senses. They can apparently detect the presence of water vapour and also differences in temperature. They also have remarkable ability to observe the passage of time. Some flowers have actively functioning nectaries only at certain periods of the day and bees learn to visit such flowers only at times when there is enough nectar to be worth getting. Forel observed in 1906 that bees came to his dinner-table looking for jam only at breakfast and tea-time. For other meals there was no jam and no bees came. They were not just attracted by the smell of jam, for when he had a jam-less breakfast or tea, the bees turned up just the same. How bees 'tell the time' is obscure. It may be partly governed by the course of their internal chemical processes, for when these are slowed down by changes in temperature or in other ways, bees can be caused to be late for an appointment.

Distance can also be measured and communicated in the bee-dance language (see p. 179). Probably what is measured is the time taken to fly to the hive from a source of food. The information transmitted may be inaccurate if the bee has a following or an adverse wind. Ribbands has suggested that hairs on a sense organ may be bent as an air-stream passes them and that the bee stores the information concerning the period during which this has happened.

Species mentioned

Apis mellifera	BA	Ay'-pis mell-if'-er-uh
— dorsata	Asia	— dor-say'-tuh
— florea	Asia	— flor'-ee-uh
— cerana subspecies indica	Asia	— serr-ay'-nuh ind'-i-cuh

CHAPTER 24

The Honey-bee: Reproduction

THE story of the reproduction of any species should really run on a circular course. In practice one must make an arbitrary decision and start either with the hen or the egg. In the case of honey bees it will be convenient to begin at a point when the hive is in a state of equilibrium: the queen is engaged in laying eggs in one cell after another; the foraging workers are collecting honey and the nurses are looking after the young brood. We shall consider in the next chapter how the whole thing is made to run smoothly by appropriate divisions of labour. Obviously, things cannot go on for ever without some change: the queen must grow old some time and the bees will also want to found new colonies. What happens will be either supersedure of the old queen by a young one or else departure of some of the bees in a swarm, or a combination of these happenings.

With simple supersedure, the queen, as she grows old or diseased, ceases to lay as many eggs as before. The workers then proceed to build new queen cells. What makes them do this has only recently been learnt. The queen bee secretes from two large glands in her head a substance which has come to be known as 'queen substance'; this has been examined by Dr. R. K. Callow and his colleagues and its chemical constitution determined; it is a fatty acid – 9-oxodec-2-enoic acid. The workers seem to need this substance and they obtain it by licking their queen. As Butler says 'they love their queen because she tastes so good'. The workers are constantly licking, touching or feeding each other and so the queen-substance is generally distributed, even to those not in direct contact with the queen. The workers do not get enough queen-substance when she is getting past her prime – perhaps also

when there are so many of them that there is not enough
of it to go round; then it is that they begin to build queen
cells and to rear new queens. Queen-substance thus acts as an
inhibitor of queen-cell building; take it away and things are
in train for producing new queens.

To accomplish this the workers have first to build extra large
cells in which to rear queens, then they have to give the
young larva the special treatment which will turn her into
a queen. Most people know that this is done by feeding her
with 'royal jelly'. The matter is, however, not as simple as
all that. It is certain that all newly hatched larvae, whatever
their ultimate destiny, are fed for a few days on the same
material, which consists largely of 'brood-food' which is a
protein-rich secretion from the glands of certain, mainly
young, worker bees. At any time up to three days after its
hatching the workers may decide to turn the young larva into
a queen, so they go on feeding her on her brood-food. If not
they provide nectar and give less brood-food and the larva
will turn into a worker. After three days the choice has been
made irrevocably and no amount of special diet added later
will turn such a larva into a queen. Another view is that the
difference in diet is a matter of quantity. A princess bee-larva
lives in a large thimble-shaped cell which is provided with
food *ad lib*. In fact, she commonly fails to finish it all before
she pupates. The workers-to-be are, on the contrary, fed by
progressive provisioning and the food given is strictly rationed:
this is necessary, for the older larvae take up so much room in
their small cells that there would be no room for food-stores
(Plate 16B). Several people have taken larvae less than three
days old and tried to turn them into queens by giving them
brood-food in unlimited quantity, but they have met with
little success. It is possible therefore that some hormone or
other substance has to be added to the food and a large cell
with lots of first-class protein composition is not enough.

When a queen, destined to supersede an older one, hatches
from her cell, the workers take no special notice of her and

she wanders around for a few days, consuming honey but doing no work. If two queens emerge together they will fight until one stings the other to death; bees are by no means immune to bee venom. Or if from a number of queen-cells one emerges first, she will fall upon the other queen cells and tear holes in them. The workers commonly complete the destruction, so that the first young queen is left in possession. Newly emerged queens emit a piping note to which other virgins will respond, even those not yet out of their cells. The first piping queen is thus helped to find potential rivals.

After a few days, if the weather is favourable, the virgin queen begins to make from one to three short flights from the hive. They serve as orienting flights to teach her geography. Queens usually mate when about ten days old with drones encountered on a flight. Mating takes place in the air and observers have seen a comet-like formation with the queen at the head and a 'tail' of pursuing drones. When a successful male succeeds in mating, his genital apparatus is extruded into the female; he becomes paralysed, falling backwards, his genitalia snap away from him and he falls dead to the ground. A female commonly mates with about five drones either on the same or on subsequent flights. Drones pass rather freely from one hive to another: a custom which doubt-less operates to make in-breeding less likely.

Drones within a colony show no interest in a nubile queen in their own hive. Only a very small proportion of perhaps a hundred produced in one hive ever succeed in mating. The rest live for four or five weeks, helping themselves to food in the hive or being fed by workers, but never gathering it for themselves. They take flights looking for a mate only on fine afternoons. At the end of the summer, any surviving drones are turned out of the hive and left to die in the cold. No wonder the word 'drone' has come to refer to lazy lay-abouts.

When the mated queen returns to the hive, she receives much more attention from workers. After a day or two workers throng around, lick her, feed her with brood-food and incidentally acquire some nice fresh queen-substance. The drones are reared in cells which are a little larger than workers' cells and have more convex dome-shaped caps. It was mentioned earlier (p. 85) that it is commonplace among the Hymenoptera for unfertilized eggs to produce males and for fertile ones to give rise to females—including, for the social species, workers. The queen honey-bee wanders round the brood-area of the hive, laying an egg whenever she finds a suitable empty cell. If she finds a worker cell she deposits a female egg and if it is a drone cell, a male egg is laid. One may well ask how on earth this is decided, for very few mistakes are made. When the queen mates, all the sperm received is stored in a special receptacle; this supply may last for years and suffice to fertilize several hundred thousand eggs. It is suggested that when a female egg is to be laid, a constricting muscle is relaxed, allowing the escape of a little sperm, while, if it is to be a drone egg, the muscle is tightened so that no sperm can get out. Control of the muscle could be determined by some stimulus set up by the shape of the cell encountered. Now, all the eggs laid have to pass to the exterior through a single passage. It is a little hard to see how this can be rendered so free from sperm after a worker egg is laid that there is no single spermatozoon left which could make unauthorized penetration into an egg intended to give rise to a drone. There is therefore some doubt as to how this practically fool-proof sex-determining mechanism really works.

It may happen that a colony, through death or otherwise, loses its queen. The workers quickly become upset and when the hive is opened give forth a 'moaning' or 'roaring' sound. This is produced through fanning of the wings of many bees and simultaneous opening of their scent glands, so that their scent is freely dispersed. In the absence of a queen, the bees soon proceed to make alterations to some worker cells, turning them into emergency queen cells, and a new queen is then reared. This is only possible if there are available young female larvae less than three days old. If there are none of these, the colony is normally bound to perish. It sometimes happens, however, that when all is apparently lost an unexplained mystery queen appears in the hive. This can probably be accounted for in the following way. Many worker bees have functioning ovaries and can lay eggs, though they are anatomically incapable of mating. Any eggs they lay must therefore produce drones. In any case they don't lay at all except when they do not get their regular supply of queen substance. It has been suggested that occasionally a worker may succeed in mating with a drone and then laying fertile eggs; but there is no evidence that this can happen. It does, however, very rarely occur that a few unfertilized eggs laid by a virgin queen or a worker will produce a female: and this is doubtless the explanation of the 'mystery queens'.

New colonies of honey bees are formed by the familiar process of swarming. Swarms may be of several kinds. In a mating swarm, one or more virgin queens, instead of going off for a private wedding and honeymoon, are accompanied by a large wedding party. The queen, or one of several, may mate and return with her party to the hive, to take over in due course from her mother. Alternatively she and her attendants may go off and found a new colony elsewhere. In the commoner kind of swarm, a 'swarming-impulse' is aroused. One view is that this is a result of overcrowding, which may have had the result that there was not enough queen-substance to go round all the members. The workers, anyway, make cells for rearing new queens and a lot of bees—swarm bees—become idle, apparently waiting for the signal to swarm. It is commonly supposed to be the old queen who goes off with the swarm; she is not the first to leave the hive, as is the case in mating swarms. A swarm will usually settle in a tree or other conspicuous place and move off again after a few hours or days to a suitable place for a new home. Very often, perhaps eight days after the 'primary swarm' has left, there may be one or more 'casts' or 'after swarms' each with a virgin queen.

Before such an event the workers will have taken care to keep any young queens apart, not allowing one to destroy her rivals, as described when supersedure was occurring.

One of the most remarkable things about swarming is the manner in which a new home is found. A few days before the swarming takes place some scout bees will have been out 'house-hunting'. These join the swarm in its temporary resting place and indicate by the dance-language to be described in chapter 26 the whereabouts of a desirable residence. Several scouts, however, may have found different possible sites. They all, somehow, 'reach agreement' for after a while all are seen to be performing a similar dance: the site of the new home is decided.

CHAPTER 25

The Honey-bee: Work in the Hive

A HEALTHY hive may contain between 20,000 and 30,000 bees, the vast majority of which are workers. The many tasks are mostly carried out through a division of labour. The cells for storing honey and pollen are mainly in the upper part of a hive; the brood cells, where the new bees are reared, are lower down. Royalty has to be kept under control: the queen will be adequately fed so long as she gets on with her task of eternally laying eggs. At peak periods she may lay 1,500 eggs, or the equivalent of her own body-weight, every day. She is given food as often as once a minute: and she must surely need it!

When a young bee first emerges from her cell, she looks wet and bedraggled but soon tidies herself up and begins to beg food from older sisters. She also takes pollen, which is protein-rich, from the stores; this helps her in production of the brood-food which she will soon be yielding. For the first few days she occupies herself in cleaning out cells and helping to keep the brood warm. Then, perhaps from the third to sixth days she will be feeding the older larvae. After that, her glands begin to produce brood-food and this she gives to the young larvae and the queen. At about a fortnight old, she begins to secrete wax, in a way to be described shortly, and to engage in comb building. Soon after that she becomes a forager and goes forth to collect nectar and pollen. Before this, however, she has made a number of short play-flights which enable her to become familiar with the surroundings of the hive. Now this is an idealized programme, by no means strictly adhered to. Worker bees in the hive are by no means hampered by restrictive practices, but will, on the contrary, often look around and carry out whatever task is most urgently needed. Foragers will, if necessary, help with household chores. Butler suggests that besides the young, household bees and the older, regular foragers, there is a category of link bees ready to perform either field or hive duties and perhaps peculiarly able to co-ordinate the two areas of work, knowing what at any one moment are the special food requirements of the colony. At the height of the summer, bees going 'all out' on their many tasks have an expectation of life of only four or five weeks. Feeding a large number of growing larvae with brood-food exhausts their food-reserves and the hard work of foraging makes them expend all the energy they have left. This is what happens to 'summer-bees'. Those emerging in the autumn have less brood to care for; their brood-food is not used up quickly and soon there is no foraging to be done; so a 'winter-bee' may live throughout the winter.

The young nurse-bees take food from incoming foragers and store it away, nectar into some cells and pollen into others. Nectar has to be converted into honey and this is done by evaporation and concentration. The hive is kept well ventilated and this greatly helps the evaporation. In addition to this, the nurse-bees process the honey, regurgitating it from their crops and exposing it on their tongues over and over again to reduce it, through evaporation, to the right consistency.

Food is placed in the larval cell where it is readily reached, rather than being given to the larva directly. It has been stated that the nurses stimulate the larvae to give up a drop of secretion, of which they partake, food exchange being mutual (see p.102). Recent observations throw much doubt on this in the case of honey-bees, though it doubtless happens in wasps. Food is, however, constantly being transmitted from one worker to another (Plate 16A). This helps to spread queen-substance around, to ensure that the scents of members of one hive are similar, and generally to weld the community into a homogeneous unit.

The wax which is used to build the combs is derived from the honey they consume and is secreted by abdominal glands; as a result of this activity scales of wax appear between the abdominal segments. The hind legs of the bees hook the scale from its little pocket on the abdomen and it is carried forward from there to the mouth. It has been calculated mathematically that the hexagonal shape of cells in the comb is that which will hold the most honey with the consumption of the least amount of wax. The walls are first made quite thick and then pared down to the requisite thinness.

Another substance used by bees in building is called propolis. This is a resinous substance obtained from a number of sources—sticky poplar buds, resin from pine trees and elsewhere. It is for sealing cracks and gaps in the hive wall. The bees apparently use anything of the right physical properties, even white lead in oil!

Another duty is that of guard bees stationed at the entrance to a hive to repel unwanted visitors. Only a few bees are needed to undertake this duty; they are perhaps most frequently younger foragers. They examine all incoming bees but are not, in the ordinary way, very particular as to whom they will admit. Bees belonging to a colony have an odour which is shared by other members of this colony and this helps guard bees to distinguish friend from stranger. The odour is partly made up of a hereditary component but depends also on what flower has been yielding the nectar which has been gathered (see p. 164) and on other inconstant factors. Differences between hive odours are evidently small. When there is a plentiful supply of honey, nectar-laden bees often 'drift' into a hive not their own, and nobody minds. When, however, food is short the guards become more critical and strangers will be turned away. Inadvertent intruders not carrying loads of nectar or pollen show signs of an 'inferiority complex'. They cringe and may offer food to the guards, but these soon attack them and hustle them away. There may also come bees of an opposite character, robbers intent on theft. These are quickly recognized by the guards because of a peculiar, hesitant, form of flight. The robber tries to escape when spotted, but if it cannot do so, there ensues a fight, one or other contestant being stung to death. Strange bees which do get into a hive presently acquire, fairly soon, enough of the odour of that hive so that they are accepted as natives.

The sting of a bee, as already mentioned, can be fatal to other bees. The venom is of unknown composition but contains protein; it comes from a mixture of the secretions of two pairs of glands. It used to be taught that bee stings were acid and wasp stings alkaline and that they could be neutralized accordingly: but that was baseless folk-lore. The bee's sting consists of two lancets, both barbed. These slide into a puncture alternately, each making fast by one of its barbs while the other goes in a little farther; and so on. The bee soon breaks away leaving the sting behind, and her death follows. A volatile substance from the bee's sting-chamber seems to excite other bees; and this explains why a bee-keeper stung by one bee is liable to be attacked by others. Bee-venom can engender some immunity, so that after receiving repeated stings a bee-keeper eventually comes to suffer less reaction. A very small proportion of people are allergic to bee-venom and suffer shock and other serious consequences when stung.

Colonies of honey-bees, in contrast to those of bumble-bees and wasps, persist as such through the winter; they can therefore get off to a, literally, flying start in the spring. This is possible because in winter the bees cluster together and their combined heat production raises the temperature within the hive to a level much above that of the world outside. A difference of $59\,^{\circ}C$ has been recorded between the temperature $(31\,^{\circ}C)$ inside and $(-28\,^{\circ}C)$ outside the hive. Usually tight clustering begins when the temperature in the hive falls to between $14\,^{\circ}C$ and $18\,^{\circ}C$. The surface area of a cluster is vastly less than that of a whole lot of separate bees and so heat-loss is much reduced by clustering. On the same principle the bees can regulate temperature by expanding and loosening the cluster, so lowering temperature, or by contracting and so raising it. It is not now thought that temperatures are kept up by purposeful movements of the bees, but rather as a result of chemical changes in their bodies related to food consumption. It will be obvious that the bees cannot keep the hive warm without fuel; and the fuel is the sugar or honey they eat. Round about February brood-rearing begins, and then, of course, food supplies are drawn upon much more freely.

The last function of bees to be discussed concerns the 'other side of the coin'; they have to prevent the hive from becoming too hot in summer. This they do by fanning. This may be just to give better ventilation, or they may bring in water, deposit it on combs and fan to increase the rate of evaporation. The evaporation of fluid as nectar is turned into honey also helps to keep temperatures down. One observer heated a hive electrically and so induced bees at the hive's entrance to start fanning. When he then sucked cool air through, the fanning abruptly stopped.

15(A) Opened nest of *Bombus lucorum*

15(B) The cuckoo bumble-bee *Psithyrus rupestris* (notice dark wings and absence of pollen-gathering apparatus on hind legs)

16(A) Worker honey-bees exchanging food

16(B) Larvae of honey-bees in their cells

CHAPTER 26

Honey Bees and Flowers

MOST flowers provide bees with both nectar and pollen. A few, such as some poppies, provide only pollen; and where male and female flowers are separate, there will obviously be some yielding nectar only. Bees also collect water and may at times obtain nectar from elsewhere than flowers—for instance the nectar-secreting glands on laurel leaves or cornflower buds. Individual foragers will concentrate largely, though rarely wholly, on either pollen or nectar. A little nectar is incorporated into pollen-loads, perhaps making it stickier and easier to transport. Most of the nectar is carried in the worker's honey-stomach and processed, as described in the last chapter, till of the right consistency and then stored in cells in the hive as honey. Pollen is automatically deposited on the bees' hairs as it moves in a flower and is often collected by very highly adapted arrangements of hairs on the hind legs. Rakes inside each leg scrape pollen off the opposite leg; it is then squeezed into a special basket on the outside of each hind leg; this is a smooth area surrounded by a fringe of hairs; there is one strong anchoring hair or spine in the middle of the smooth area. When the worker returns to the hive she uses her middle legs to push off the pollen. Household bees then push it properly into the appropriate cells.

Some flowers have colour-contrasts, called nectar-guides, believed to lead the bee to where the nectaries are; there may also be slightly different or stronger scents in such areas. Some flowers are fairly intricate and the way to the nectar-stores may take a little learning. That is probably, as Darwin pointed out, the reason why bees tend to concentrate on one kind of flower at a time: they do not have to keep learning new tricks for finding nectar; they may keep to one flower crop for days on end if it is yielding well. More than that, they often keep to a comparatively small foraging area a few yards across, or to a particular bush. All this has been discovered, of course, by watching marked bees. When they visit a dish of syrup, one bee may alight again and again on the same place on the dish. This habit is very convenient for the nurserymen who wants to get pure seed of a particular variety of flower, for if they always visit the same spot he will get little mixture from other strains in his garden. It is less convenient for the growth of apples, many of which are self-sterile and have to be crossed with other varieties. Some flowers yield most of their nectar at particular times of day and, if these times are different for two flowers, a bee may divide her time between the two, so as to get the best possible nectar-harvest. Honey, as mentioned earlier, contains mainly sugars, though heather honey has rather more protein than other kinds: in pollen there is far more protein. It is suggested that one advantage of food-sharing between worker-bees is that they learn in this way which kind of food is most needed for feeding the brood.

New sources of supply of food are being regularly discovered through the activities of comparatively small numbers of scout-bees. These have a characteristic flight a foot or two above the ground, always investigating bright objects which might be what they are seeking. Several workers have produced evidence that colour attracts a bee first and that only if it has a promising scent will the matter be followed up. Only after several journeys home with a load of provisions will the scout perform the dances, to be described in a moment, which will guide other workers to the food-source. There does not exist a permanent staff of scouts: various bees may become scouts for a time and a scout who has found something really worth while will often become a regular forager and so exploit her discovery. Scouts can apparently become accustomed to seeking flowers of a particular kind. Some which had been 'trained' on white mustard flowers were taken, along with their hives, to a heather moor. Three weeks later they were found to be bringing in pollen from mustard growing a mile away, while bees from adjacent hives having no mustard-training, were not getting that kind of provender at all.

There is a good deal of evidence that scent plays a large part in their activity, more in fact than colour does. It may be desirable to attract bees to a particular crop, either because it is a good source of nectar for them or because the bees are wanted for cross-pollinating the flowers and so getting a good yield of seed. One method which has been tried is to offer the bees syrup in bowls garlanded with flowers of a particular kind, say red clover. Another way is to apply the scent inside the hive. In trials with the red clover, flax, beans and other crops, there have been claims for definite advantages from direction-by-scent, either in greater yield of honey or of better setting of seed in the crops; there is now some doubt as to the validity of the claims.

We come finally to perhaps the most remarkable discovery in the whole field of wasps and bees, that of the dance-language of honey-bees, which directs them to sources of food. A number of people had felt sure that bees had some means of telling their colleagues where to go to a good crop; and in 1920 K. von Frisch in Munich described how the bees 'communicated' by means of dances.

Scouts or successful foragers when they return to the hive perform two kinds of dances, a round dance and a tail-wagging dance. Von Frisch thought at first that the former was used to recruit foragers to nectar-sources, while the latter referred to sources of pollen. In 1946 he revised this view. The round dance is performed when the source is roughly 100 yards from the hive or less. The bee rotates on one spot, clockwise for a while, then 'widdershins' or anticlockwise. The dance is thus like a figure of eight with the two loops almost superimposed. As the distance of the nectar-source from the hive increases, its form changes through a transitional sickle-shape,

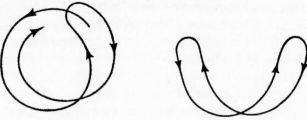

Round and sickle dances. Source of food 11 yards away

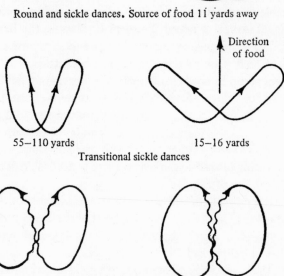

55—110 yards   15—16 yards

Transitional sickle dances

110 yards. Figure of eight dances

15. Dances of honey-bees

as shown in Fig. 15 to a figure of eight. On the middle part of
the figure of eight, the bee waggles its tail, first signs of a
waggle being visible at 28 yards between hive and food-source.
Then as that distance increases, the length of that vertical
run increases and the number of waggles per run increases;
the latter change is more regular than the former. At the same
time the number of waggle-runs per inch decreases (see Fig.
15). The human eye cannot follow the waggles fast enough to
count them accurately; the bees manage, however, to do this.

Thus the bees convey to their comrades an idea of distance. It is not, however, strictly distance which is measured, but rather, as mentioned on p. 165, the time taken for the bee to fly *from* the hive *to* the food-source. This will be greater if there is a contrary wind or if the source lies uphill from the hive. The direction of the source is also conveyed, particularly in the figure of eight dances by the way the middle, straight, part of the figure points. It would seem simple for it to point in the direction in which the source lies. The thing is, however, much more complicated than that. First of all, the dance is normally performed, not on a horizontal surface but on the vertical face of the comb; and, second, the information given is related to the position of the sun at the time. If the straight run is vertically up, the feeding place lies in the same direction as the sun, if vertically down it is away from the sun. If the angle of the sun is to left or right, then the food-source is at a corresponding angle away from the sun. The dancing bee thus translates stimuli from light into terms of gravity and the other workers translate from gravity back into light. Now the sun does not 'stand still in the sky' and the dancers and their observers take cognizance of this. The dancers change the angle of their dance with the lapse of time by the same value as the apparent angular movement of the sun in the sky.

The dancing is apparently instinctive behaviour, for bees which have never seen others dancing know how to do it. Butler thinks it unlikely that they are purposefully conveying information. On the other hand, young bees have apparently to watch dances for several days before they are able to interpret them.

Occasionally the dances are carried out on a horizontal surface, the alighting board in front of the hive. The straight run then points directly at the source of food. The dancers manage this all right as long as they can see the sun or even a little bit of the blue sky; this is because of their ability, mentioned before (p. 163) to detect the plane of polarization of light. Not only sunlight but light from the blue sky is polarized, that is the light waves vibrate in only one plane instead of in all planes at once. In some experiments, however, von Frisch found that bees could 'calculate' their angles correctly even when the sky was wholly covered by cloud; this he could not explain.

Information imparted by dancing is supplemented by the use of scent. Foragers which have recently danced have been seen to fly round the feeding place with scent-glands exposed, thus helping to guide new foragers to this profitable place.

One may well ask how the bee-language can have arisen in the course of evolution. Butler has observed the behaviour of other *Apis* species in Ceylon. *Apis cerana* subspecies *indica* carries out both round and wag-tail dances like those of *A. mellifera*. The little honey-bee *A. florea* has a single comb the top of which can serve as a horizontal dance-floor. On this, these bees perform both round and tail-wagging dances, the straight run of the latter pointing directly to the source of food. They never dance on the vertical face of the comb; so it is likely that the vertical dance has been derived from a more primitive horizontal one. The giant honey-bee, *A. dorsata* does dance on its comb's vertical face but nevertheless orients its dances to the sun rather than to gravity.

CHAPTER 27

The Evolution of Wasps and Bees

NUMEROUS references have been made in the preceding
chapters to aspects of the lives of wasps and bees having
bearing on their evolution. This last chapter will try to look
at this evolution in perspective. It can be considered under
four headings: the origin of aculeates from parasitic hymenoptera, the development of social life, the re-acquisition of
parasitic behaviour by some species and genera, and finally
the development of more flexible nervous responses.

Most of the changes to be discussed must have been taking
place over millions of years. The earliest fossil sawflies have
been found in remains of the early mesozoic period; they were
thus contemporaries of the dinosaurs. Parasitic hymenoptera
are known from the late mesozoic and wasps from the tertiary
period, probably thirty million years ago. Mention has already
been made of the probable origin of most solitary and social
wasps, from the superfamily *Scolioidea*, which includes many
of the wandering wasps described in chapter 11. It was there
noted that a wasp did not really differ in its habits from a
member of the orthodox parasitic hymenoptera when it laid
its egg on its victim and left it to develop at the spot where the
prey was found, building no nest for it. The resemblance is
still closer when the attacked host recovers from a temporary
paralysis and for a while carries on a normal existence, as do
spiders attacked by *Homonotus* (p. 19).[59] Chapter 15 dealt
with the evolutionary steps which led from this type of
behaviour to that seen in the great bulk of solitary wasps. W.
M. Wheeler[61] discussed the evolution of parasitic aculeata in
a fascinating way about fifty years ago. He wrote that the wasps
related to *Scolia* and *Mutilla* 'occupy a position, ethologically
as well as structurally, midway between the higher wasps and

the *Ichneumonidae*. The higher wasps in constructing nests and provisioning them with paralysed insects merely elaborate the same fundamental behaviouristic theme or pattern, the main features of which were also retained by the solitary bees even after they had ceased to capture insect prey and had become pollenivorous and nectarivorous.'

A number of genera of wasps and bees have independently developed traits which may be regarded as steps on the road to social life. A first step would presumably be the tendency of 'solitary' species to build many nests close together: this might be because of a shortage of suitable nesting sites, or of some benefit gained by mutual protection against enemies, or merely through innate conservatism leading a new generation to nest on its home ground. Another step is perceived where we find some wasps related to *Crabro* and some Andrenas using a common entrance burrow from which their individual nesting burrows branch off. Still more important is the acquisition of the habit of sometimes, and later always, provisioning progressively. This helps to build a social instinct for the mother has some contact with her developing offspring. Without it she never sees her descendants except in the form of an egg. It may be associated, as in *Bembix*, with provision of dead instead of paralysed prey or even (p. 90) of chewed-up prey. Learning to take care of several nests simultaneously, as *Ammophila pubescens* and a tropical *Stenogaster* do, is a development which sees to it that progressive provisioning does not involve spending several days looking after one larva.

Three genera as far apart as *Polistes*, *Halictus* and *Bombus* have independently reached what may be regarded as truly social life. In each case it was a crucial step when young females began to help their mother to rear the next generation. This must soon have led to their ceasing to be fully fertile and thus becoming workers. The stages in this development are best seen in the *Halicti*. There is presumably much importance to be attached here to the ability of a fully fertile female, either the mother in the case of *Halictus* and *Bombus* or a dominant queen in the case of *Polistes*, to suppress other females and inhibit the development of their reproductive organs. In *Bombus*, and even more in *Apis*, this has gone so far that the workers are usually considerably smaller than their queen. In *Apis* the difference is established by the giving of brood-food to queen larvae and the associated social behaviour is regulated through queen-substance. *Apis* queens differ also from the queens of other social species in that they do not at any time take part in foraging activities. *Apis* is also the only genus which has managed to keep its colonies going perenially in temperate climates, though species of *Polistes* and *Bombus* may achieve this in the tropics. With the development of a dance-language and other refinements of behaviour, honey-bees have clearly advanced far beyond any other wasps or bees.

Wheeler[61] has discussed at some length the bees and wasps which have reverted to a parasitic life. He goes so far as to suggest that this may be easier for them because 'the ancient parasitoid habits of the Ichneumonid ancestry still abide as a latent memory . . . in the constitution of the whole aculeate group'. Between 15 and 20%, representing fully seventy genera of bees, have become parasitic; among wasps the habit is less widespread. Adoption of a parasitic life has led, probably independently, to structural changes in the genera concerned. Their larvae, at least in the earlier stages of those which kill the host larva themselves, have sharp sickle-shaped jaws and better-than-usual powers of locomotion. Adults have the hard smooth skins necessary for defence. They are also commonly more brightly coloured than their hosts, perhaps because they are more exposed to attack by birds and other predators. They have been referred to in this book as cuckoos, but the analogy with the cuckoo is not exact: the young cuckoo disposes of its hosts' eggs or fledglings but does not actually eat them as many larval cuckoo-wasps do.

The parasitic habit can have arisen very early, for many aculeates, not normally parasitic, are apt to steal the food of other individuals and this may happen more especially when the food-supply is short. A very strong stimulus may be applied when there is urgency for oviposition. This is apparent especially in parasitic aculeates, which, when their egg-laying is frustrated, may lay several eggs in one cell. This is, of course, a useless act: only one larva will survive anyway. Wheeler writes: 'When the parasitic habit is once started it tends necessarily, owing to the saving of energy which would otherwise be expended in work, to accelerate the maturation of ova and thus to become more and more confirmed by one of the circular processes so familiar to the physiologist.' One may perhaps add to this that as a result of counter-measures by the host, the parasite may in the end have to do much work in countering the counter-measures.

Most parasitic aculeates closely resemble their hosts and have obviously been derived from them; just how recently was discussed in chapter 22 dealing with *Bombus* and *Psithyrus*. There are even species in a genus, parasitic on other members of the same genera. Examples have been mentioned as occurring in the genera *Pompilus*, *Polistes* and *Vespula*. Their origin is easy to understand. More difficult are the instances where the parasitic and host genera are dissimilar, as are *Nomada* and *Andrena*, and, even more obviously *Epeolus* and *Colletes*. To quote Wheeler again: 'We must assume that in some cases the primitive host genera are now extinct, that in some cases, therefore, the parasites have come to infest species of genera to which they have no morphological affinity, that many parasites are directly derived from other parasitic genera . . .' Further difficulty in elucidating the origins of parasites arises from a natural tendency of all parasites to acquire, by convergent evolution, similar characters convenient for their particular kind of existence. In this connection Wheeler points out that the hosts of parasites mostly belong to big dominant genera such as *Andrena*, *Halictus*, *Osmia* and *Bombus* among the bees. 'Probably', he writes, 'the dominant genera, owing to their abundance in individuals and the wide distribution of their species, would act like great nets set to capture any parasites which have overstepped the bounds of good parasitic manners by bringing their original host species to the verge of extinction.' He gives a list of some forty genera of parasitic bees and their hosts, and in a parallel column the genera from which he surmises that the parasites have been derived. In twenty-five out of forty instances the parasitic genus is believed to have originated in that of the host.

It was formerly fashionable to contrast instinct with intelligence, regarding the two things in terms of black-and-white differences. Many examples have been given in this book of instinctive behaviour of a rigidity which has, in our eyes, ridiculous consequences. One may refer to Fabre's *Sphex* (p. 29) which, when he interfered with it, uselessly repeated an instinctive act forty times over. The many observations of others on various wasps and bees give further proof that too rigid an instinct may lead to futile behaviour: the instinct works perfectly as long as things are normal; it too often fails in the face of unforeseen difficulties. There is plentiful evidence that the insect's actions depend upon a chain of reflexes: only when one action has been completed, does it proceed to the next action. Tinbergen[25] points out that *Philanthus* has to be 'set' by one stimulus before a second one will have the proper effect. In catching bees it is 'set' by the visual stimulus of an object resembling the desired prey; only after that does reception of the appropriate scent trigger off the predatory leap. But if the scent of a bee is supplied without the visual stimulus, the wasp is quite unmoved. In contrast to this stereotyped behaviour, it seems that when the wasp is struggling with the bee, trying to apply its sting in a particular vital spot, it must be able to respond in quite a complicated way to unforeseeable contortions of its victim. Some of Fabre's experiments on *Chalicodoma* afford other good instances (p. 140).

We cannot help marvelling at how complicated stereotyped behaviour can become. *Ammophila pubescens* (p. 37) visits several nests being simultaneously provisioned and carries away with it a memory of what each nest requires. Many wasps and bees are able to commit to memory the essential features of the surroundings of their nests after a locality survey of only a few seconds. While we are impressed with the foolish results of a rigid instinct, we cannot fail to be interested also in the many instances in which the insect *does* succeed in adapting its behaviour to changed circumstances. Not all Fabre's *Sphex* behaved as stupidly as the one just quoted: some learnt how not to be fooled any longer. Honey- and bumble-bees are adaptable and can be trained to visit particular flowers or trays of syrup marked with particular colours and shapes. It is useless to discuss at what stage flexibility of reaction to stimuli justifies us in talking about intelligence instead of instinct.

Most amateur entomologists in Britain and elsewhere are attracted to make collections and to some extent study the habits of those conspicuous insects, butterflies and moths. Very few pay any attention to the many species of wasps and bees. Yet, as I hope this book has revealed, their life histories are so varied and their habits so remarkable that they deserve far more attention than they ever receive; and there is very much about them yet to be learnt.

Further Reading

The following books on the habits of wasps and bees are referred to by numbers in the preceding pages:

1. Butler, C. G. *The world of the honey-bee.* Collins, 1954
2. Carthy, J. D. *Animal navigation.* Allen and Unwin, London, 1956.
3. Duncan, C. D. *The biology of North American Vespine wasps.* Stanford University Press, 1939.
4. Evans, H. E. *Wasp farm.* Harrap, 1964.
5. Evans, H. E. *The comparative ethology and evolution of the sand wasps.* Harvard University Press, 1966.
6. Fabre, J. H. *The wonders of instinct.* Fisher Unwin.
7. Fabre, J. H. *The life of the fly.* (Translation by Teixeira de Mattos) Hodder and Stoughton, 1913
8. Fabre, J. H. *The mason bees.* (Translation by Teixeira de Mattos) Hodder and Stoughton, 1914.
9. Fabre, J. H. *Bramble-bees and others.* (Translation by Teixeira de Mattos) Hodder and Stoughton, 1915.
10. Fabre, J. H. *The hunting wasps.* (Translation by Teixeira de Mattos) Hodder and Stoughton, 1916.
11. Fabre, J. H. *More hunting wasps.* Dodd, Mead & Co. New York.
12. Free, J. B. and Butler, C. G. *Bumble bees.* Collins, 1959.
13. Krombein, K. V. *Trap-nesting bees and wasps. Life histories, nests and associates.* Smithsonian Press, Washington, D.C., 1967.
14. Michener, C. D. and Michener, M. H. *Anerican social insects.* Van Nostrand (Toronto, New York and London) 1951.
15. Olberg, G. *Das Verhalten der Solitare waspen Mitteleuropas.* Veb. Deutscher Verlag der wissenschaften 1939. (Legends to plates in English.)
16. Peckham, G. W. and Peckham, E. G. *Wasps, social and solitary.* Constable, 1905.
17. Rau, P. and Rau, N. *Wasp studies afield.* Princeton University Press, 1918.

18. Reinhard, E. G. *The witchery of wasps.* Century Co. New York and London, 1926.
19. Ribbands, C. R. *The behaviour and social life of honey-bees.* Bee Research Association Ltd., London, 1953.
20. Richards, O. W. *The social insects.* Harper, London, 1953.
21. Sakagami, S. F. and Michener, C. D. *The nest architecture of sweat-bees* (Halictinae). University of Kansas Press, Lawrence, 1962.
22. Shafer, G. D. *The ways of a mud-dauber.* Stanford University Press.
23. Sladen, F. W. L. *The bumble-bee: its life and how to domesticate it.* Macmillan, 1912.
24. Tinbergen, N. *The study of instinct.* Oxford University Press, 1951.
25. Tinbergen, N. *Curious Naturalists.* Country Life Ltd., 1958.
26. Williams, C. B. *Insect migration.* Collins, 1958.

Books for the identification of wasps and bees are far from satisfactory. The standard work for Britain is that by E. Saunders: *The Hymenoptera aculeata of the British Islands* (Reeve, London, 1896). It is out of date and out of print. Four Volumes on bees, wasps, chrysids and ants, respectively, are in preparation in the Royal Entomological Society's series of Handbooks for the identification of British Insects, and these should be excellent when they appear. Meanwhile, some information can be gathered from *Bees, wasps, ants and allied insects* by E. Step (Warne, 1932): a new edition is in preparation.

For North and Central Europe there is: *Die Hymenopteren Nord-und Mitteleuropas* by O. Schmiedeknecht (Fischer, Jena, 1930).

For Eastern North America we have: *Bees of the Eastern United States*, in two volumes by T. B. Mitchell. (Carol. agric. exp. station. techn. bulletin 141.)

Articles in journals referred to by numbers in the text.

27. Cooper, K. W. 'Venereal transmission of mites by wasps . . . *Trans. Am. Ent. Soc.*, 1953, 79, 13.
28. Cooper, K. W. 'Biology of Eumenine wasps. *V.* Digital communication in wasps.' *J. Exp. Zool.* 1957, 134, 469.

29. Evans, H. E. 'Comparative ethology and the systematics of spider wasps.' *Syst. Zool.*, 1953, 2, 155.

30. Evans, H. E. 'Ethological studies of digger wasps of the genus *Astata*.' *Jl. N.Y. Ent., Soc.*, 1957, 65, 159.

31. Evans, H. E. 'A study of *Bembix U-scripta*, a crepuscular digger wasp.' *Psyche Camb.*, 1960, 67, 45.

32. Evans, H. E. 'The evolution of prey-carrying mechanisms in wasps.' *Evolution Lancaster, Pa.*, 1962, 16, 468.

33. Evans, H. E. 'Observations on the nesting behaviour of *Moniaecera asperata* (Fox).' *Insectes soc.*, 1964, 11, 71.

34. Evans, H. E. 'Notes on the nesting behaviour of *Philanthus lepidus* Cresson.' *Psyche Camb.*, 1964, 71, 142.

35. Evans, H. E. 'The classification and evolution of digger wasps as suggested by larval characters.' *Ent. News* 1964., 75, 225.

36. Evans, H. E. 'Simultaneous care of more than one nest by *Ammophila azteca* Cameron.' *Psyche Camb.*, 1965, 72, 8.

37. Evans, H. E. 'The accessory burrows of digger wasps.' *Science N.Y.*, 1966., 152, 465.

38. Evans, H. E. 'The behaviour patterns of solitary wasps.' *Ann. Rev. Ent.* 1966, 11, 123.

39. Evans, H. E. and Gillaspy, J. E. 'Observations on the ethology of digger wasps of the genus *Steniolia*.' *Am. Mid. Nat. Monogr.*, 1964, 72, 257.

40. Evans, H. E. and Lin, C. S. 'Biological observations on digger wasps of the genus *Philanthus*.' *Wasmann J. Biol.*, 1959, 17, 115.

41. Ferton, C. H. 'Nouvelles observations sur l'instinct des pompilides.' *Actes de la Soc. Linn de Bordeaux* 1897., 52, 3.

42. Ferton, C. H. 'Notes detachees sur l'instinct des hymenopteres melliferes et ravisseurs.' *Ann. de la Soc. Ent. de France.*, 1901, 70, 83.

43. Ferton, C. H. 'Ibid.' 1905, 74, 56.

44. Ferton, C. H. 'Ibid.' 1908, 77, 535.

45. Ferton, C. H. 'Ibid.' 1909, 78, 401.

46. Ferton, C. H. 'Ibid.' 1910, 79, 145.

47. Ferton, C. H. 'Ibid.' 1911, 80, 351.

48. Ferton, C. H. 'Ibid.' 1914, 83, 81.

49. Hamm, A. H. and Richards, O. W. 'Biology of British *Crabronidae*.' *Trans. Ent. Soc. Lond.* 1926., 74, 297.

50. Hamm, A. H. and Richards, O. W. 'Biology of British fossorial wasps.' *Trans. Ent. Soc. Lond.* 1930., 78, 95.

51. Hartman, C. 'Observations on the habits of some solitary wasps in Texas.' *Bull. Univ. Tex. scient. ser.* No. 6., 1905.

52. Iwata, K. 'Comparative studies on the habits of solitary wasps.' *Tenthredo* 1942., 4, 1.

53. Malyshev, S. I. 'The nesting habits of solitary bees.' *EOS. Revista Española de Entomologia.*, 11, 201.

54. Pagden, H. T. 'Observations on the habits and parthenogenesis of *Methoca ichneumonoides* Latr.' *Trans. ent. Soc. Lond.* 1926. 74, 591.

55. Perkins, R. C. L. 'The British species of *Andrena* and *Nomada*.' *Trans. ent. Soc. Lond.* 1919., 218.

56. Perkins, R. C. L. 'The assembling and pairing of *Stylops*.' *Entomologists mon. Mag.* (1918), 4, 129.

57. Richards, O. W. 'The specific characters of the British bumble-bees.' *Tr. Ent. Soc. Lond.* 1927., 75, 233.

58. Richards, O. W. 'A study of the British species of *Epeolus* Latr. and their races.' *Trans. Soc. Brit. Ent.* 1937., 4, 89.

59. Richards, O. W. and Hamm, A. H. 'The biology of the British Pompilidae.' *Trans. Soc. Brit. Ent.* 1939., 6, 51.

60. Spooner, G. M. 'The British species of Psenine wasps.' *Trans. Roy. Ent. Soc. Lond.* 1948., 99, 129.

61. Wheeler, W. M. 'The parasitic aculeata, a study in evolution.' *Proc. Amer. Phil. Soc.* 1919., 58, 1.

Classification

Superfamily   **SCOLIOIDEA**
    Family   Scoliidae
    Genus   Scolia
    Family   Tiphiidae
    Genus   Tiphia
    Family   Methocidae
    Genus   Methoca
    Family   Mutillidae
    Genus   Mutilla
    Family   Sapygidae
    Genus   Sapyga

Superfamily   **BETHYLOIDEA**
    Family   Chrysididae
    Genera   Chrysis
                                           Chrysura
                                         Tetrachrysis
    Family   Cleptidae
    Genus   Cleptes

Superfamily   **VESPOIDEA**
    Family   Vespidae
    Sub-family   Eumeninae
    Genera   Eumenes
                                           Ancistrocerus
                                         Oplomerus
                                         Stenogaster
                                         Synagris
                                         Odynerus
                                         Monobia
    Sub-family   Vespinae
    Genera   Vespa

Superfamily **VESPOIDEA** (*cont.*)
   Genera
                                      Vespula
                                        Dolichovespula
                                        Polistes

Superfamily **POMPILOIDEA**
   Family   Pompilidae
   Genera
                                        Pompilus
                                        Anoplius
                                        Pompiloides
                                        Homonotus
                                        Dipogon
                                        Ceropales
                                        Evagetes

Superfamily **SPHECOIDEA**
   Family   Sphecidae
   Sub-family Sphecinae
   Genera
                                        Sphex
                                      Priononyx
                                      Podalonia
                                      Ammophila
                                      Sceliphron
                                      Chalybion
                                      Isodontia
   Sub-family Mellinae
   Genus   Mellinus
   Sub-family Pemphredoninae
   Genera
                                        Passaloecus
                                        Psen
                                        Psenulus
                                        Mimesa
   Sub-family Astatinae
   Genus   Astata
   Sub-family Larrinae
   Tribe   Trypoxylini
   Genera
                                        Trypoxylon

Superfamily SPHECOIDEA (*cont.*)

		Trypargilum
Tribe	Crabronini	
Genera		Lindenius
		Crabro
		Crossocerus
		Moniaecera
Tribe	Oxybelini	
Genus		Oxybelus
Sub-family	Philanthinae	
Genera		Philanthus
		Cerceris
		Aphilanthops
		Clypeadon
Sub-family	Nyssoninae	
Genera		Nysson
		Gorytes
		Sphecius
		Bembix
		Bembicinus
		Microbembex
		Stizus
		Steniolia

Superfamily APOIDEA

Family	Colletidae	
Sub-family	Colletinae	
Genus		Colletes
Sub-family	Hylaeinae	
Genus		Hylaeus
Family	Andrenidae	
Genus		Andrena
Family	Halictidae	
Genera		Halictus
		Sphecodes
Family	Melittidae	

Superfamily APOIDEA (*cont.*)

Sub-family	Melittinae	
Genera		Melitta
		Pseudocilissa
Sub-family	Dasypodinae	
Genus		Dasypoda
Sub-family	Megachilinae	
Genera		Megachile
		Coelioxys
		Osmia
		Anthidium
		Dianthidium
		Stelis
		Dioxys
		Chalicodoma
Family	Apidae	
Sub-family	Anthophorinae	
Genera		Anthophora
		Melecta
		Eucera
		Epeolus
		Nomada
Sub-family	Xylocopinae	
Genus		Xylocopa
Sub-family	Apinae	
Genera		Bombus
		Psithyrus
		Apis

Index

Index

Due